HUMAN BIOLOGY LAB BOOK

Third Edition

Michael B. Clark

and

Michael R. Riddle

Suspended Animations, **Publisher**
Jamul, CA

HUMAN
BIOLOGY LAB BOOK

Third Edition

Illustrations: Sandra Schiefer, © 1994, 1995, 1996, & 2006, *Schiefer Enterprises*

Layout Design, Editing: Sandra Schiefer

Editing, Electronic Typesetting and Layout:

> Lily Splane
> *Anaphase II Publishing*
> 2739 Wightman Street
> San Diego, CA 92104

This book is printed by:

> *Commercial Printing Centre*
> 1585 N. Cuyamaca Street
> El Cajon, CA 92020

ISBN 978-1-885380-62-3

Printed in the United States of America by:

Suspended Animations
20275 Deerhorn Valley Road
Jamul, CA 91935

January 2007

Foreword

Everyone likes to look through **Human Biology Lab Book**. And why not, it looks like fun! But it's more than student-friendly. This manual covers all of the topics expected in a traditional lab class for non-majors, and yet is clear, direct, and easy to understand. Furthermore, the labs have been classroom tested by over 50,000 students. The changes suggested by those students, their teachers, and the lab technicians have been incorporated into the third edition of **Human Biology Lab Book**.

Our goal is to provide basic labs that are self-contained and totally dependable. Other lab manuals usually require that the organizational work be handled by the teacher and the lab technician. They provide only a skeletal set of instructions and guiding questions, leaving most of the technical problems and student questions unaddressed. The teacher and the lab technician are in service to these books. In our opinion, that idea is exactly backwards. A good book should serve the students, the lab technician, and the teacher who adopts it. **Human Biology Lab Book** and its supplementary materials were created with that in mind. Because our labs are easy to understand, students can become self-directed. The open-ended format of each lab activity allows teachers to expand a favorite topic and add their own experiments to the lab without disrupting the overall organization.

The lab technician will be pleased to discover that our labs are easy to set up and maintain during the week. All experiments are written with supply budgets and safety in mind. The labs contain reminders to the students about proper lab procedures. However, those statements are not meant to replace the thorough discussions of safe lab behavior and proper equipment handling that are a necessary part of every lab class.

Finally, **Human Biology Lab Book** owes its success to many people. We would like to thank the teachers and students and lab technicians at all the schools who have used our lab books. And we offer a very special thank you to those on the following campuses whose comments and suggestions have helped us to make **Human Biology Lab Book** better every year.

Central Arizona College, AZ	IVY Tech State C.-Valparaiso, IN	Jefferson Community College, NY
Paradise Valley Comm. Coll., AZ	Carroll Comm. College, MD	Lane Community College, OR
Pima C. Coll.-Desert Vista, AZ	Oakland Comm.College, MI	SW Oregon Comm. College, OR
Pima C. Coll.-West, AZ	Southwestern Michigan Coll., MI	Butler County Comm. College, PA
College of the Canyons, CA	Augsburg College, MN	Geneva College, PA
Laney College, CA	Hibbing Comm. College, MN	Texas Woman's University, TX
Mission College, CA	Inver Hills Comm. College, MN	Tomball College, TX
Ohlone Comm. College, CA	Itasca Comm. College, MN	Tomball-Willow Chase, TX
San Jose City College, CA	MSU Coll. of Technology, MT	Kearns High School, UT
Santa Rose Junior College, CA	New Mexico State Univ., NM	Murray High School, UT
Sierra College, CA	San Juan College, NM	Salt Lake Community College, UT
Southwestern College, CA	Central Piedmont Comm. Coll., NC	Saint Martin's College, WA
Marycrest Intl. University, IA	Rowan-Cabarrus Comm. Coll., NC	South Seattle Comm. Coll., WA
College of Southern Idaho, ID	Camden County College, NJ	Whitworth College, WA
Lincoln Trail College, IL	Warren County Comm. College, NJ	Alderson-Broaddus College, WV
IVY Tech C. C.-Sellersburg, IN	Hofstra University, NY	West Liberty State College, WV

Suspended Animations, Publisher, 20275 Deerhorn Valley Road, Jamul CA 91935

Notes To The Student
Welcome to *Human Biology Lab Book*

This book was written for you—the non-major biology student. *Human Biology Lab Book* uses an approach that is easy to understand. It allows you to build your knowledge of science one step at a time, no matter what your previous background.

A lab class is quite different from a lecture class. Mostly, it is more "doing" than listening, and more "looking" than taking notes. Some students are apprehensive because of these differences, and they worry about their chances of success. The comments and suggestions below should answer most of your questions about how to succeed in your lab class.

1. We use words sparingly, constructing each sentence as a clue to something important about the topic. Read carefully, and work step-by-step through the instructions and explanations.

2. Always practice safety in the lab classroom. Make sure that you understand the instructions for properly using lab equipment, lab materials, and chemical solutions. Ask your instructor when you have a question or any confusion about safe lab procedures.

3. Be prepared. Briefly read the lab activities *before* you come to class. If you and your lab partner are prepared, you will learn more during the lab and will finish the work before the end of class.

4. Pay particular attention to the ***bold-italics*** terms and concepts. These designate the central themes and definitions of words used in the lab. They are important ideas and many will be found on your lab tests.

5. Take time to review your work. Most students leave as soon as they finish the lab. Later, they are surprised when their answers are incorrect. Take advantage of the last half-hour of each lab. It is an excellent opportunity for checking your understanding of the topic with your instructor and other students.

6. Your instructor will have more time to give study hints near the end of the lab class when the less interested students have already left. This is the perfect opportunity for you to ask what will be emphasized on the lab test.

7. Come early to the first day of lab class. Check out the other students. One (or more) of them is going to be your lab partner. Pick a good one! Some students are not interested in learning biology, and will make you do all the work. Find someone who is serious about success. Be a good partner for them. Look for several other students who are good lab partners and form a study group for tests.

Finally, have fun! Taking an active interest in the lab activities will make the time pass quickly, and will increase your chances for success in biology lab class. Good luck to you, and let us know how *Human Biology Lab Book* has helped you. We welcome your comments and suggestions.

Suspended Animations, Publisher

HBL Table Of Contents

MEASUREMENT

Summary Questions

1. There are two kinds of fractions. Name them and give an example of each.

2. Define percent.

3. Why is it important to understand the metric system of measurement?

4. What is the significance of the decimal point in metric calculations?

5. Define centi, milli, and kilo.

6. What are the metric standards for measuring length, volume, and weight?

7. What are the English equivalents of measurement for the above metric standards?

8. What is your height and weight in metric measurement?

MEASUREMENT

INTRODUCTION

Some people insist that measuring things is the only way to get a fair deal from another human being. But it is also said that the measuring of things is at the heart of all mistrust between people. Whatever the correct judgment may be, historians tell us that measurement systems are based on political and economic needs, and can be found wherever humans have a network of dealing with each other.

Originally, the units of measurement depended on the type of material being exchanged. For example, a farmer selling apples would price them by the cart-load, not by the bucket. But when selling milk, a cart-load of liquid would have been ridiculous, and a bucket more appropriate. To prevent squabbles among merchants and buyers, *standard* "cart-loads" and "buckets" were determined, and these became the basis for the systems of measurement we use today.

As science developed in civilized countries, the need for scientific measurement began to conflict with the measurement systems used for trade. There are problems with adapting the common system to a science system. In grade school, children are taught the English System which uses inches, feet, yards; cups, pints, quarts, gallons; ounces, pounds, tons. However, a quart in Canada is equal to 1.136 liters. A quart in the United States is equal to 0 946 liters. And in Mexico there are no quarts, only liters. For trade purposes, these inequities can be worked out, but in science or industry where precision is important, the English System simply will not do.

Then, 100 years ago, the International Metric System was devised to standardize measurement around the world. This system provides exact precision in *powers of 10*. There is a standard reference unit used in each measurement category. On either side of the reference unit, the units increase by 10, 100, and 1000 or decrease by $\frac{1}{10}$, $\frac{1}{100}$, and $\frac{1}{1000}$.

The universal standard unit for weight is the *gram* (g). The standard for volume is the *liter* (l). The standard unit for length is the *meter* (m).

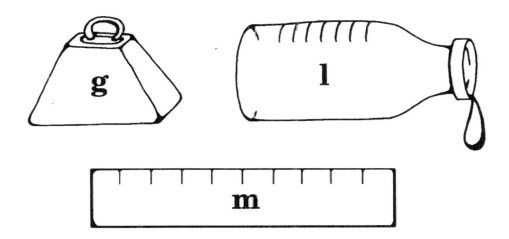

Science embraced the Metric System and so did the majority of the countries of the world. It was anticipated that the United States would be converted to the Metric System by now. It is unfortunate that so much confusion is created by converting quarts to liters, pounds to grams, and inches to meters. The solution is familiarity and practice until the United States completely adopts the Metric System.

In today's lab you will be introduced to the Metric System. You will learn to use some simple measuring equipment that will help you remember the different units of the Metric System.

ACTIVITIES

ACTIVITY #1

"SOME BASICS OF THE METRIC SYSTEM"

REVIEW

FRACTIONS

Fractions represent parts of a whole, and there are two methods of expressing them:

> *First*—Simple Fractions
> Examples are $\frac{1}{4}$, $\frac{1}{2}$, $\frac{9}{25}$, etc.

> *Second*—Decimal Fractions
> Examples are 0.25, 0.50, 0.36, etc.

FRACTION CONVERSION

 RULE 1 — *If you want to convert a simple fraction into a decimal fraction, then divide the top number of the simple fraction by the bottom number.* Therefore, the simple fraction $\frac{9}{25}$ is converted into decimal form by:

$$25\overline{\smash{\big)}\ 9.00} \quad \rightarrow \quad 0.36$$

 RULE 2 — *If you must convert a decimal fraction into a simple fraction, then put the decimal fraction over 1.00 (this represents the whole), and reduce the top and bottom numbers to the simplest fraction.* Therefore, the decimal fraction 0.36 converts into:

$$\frac{0.36}{1.00} \quad \text{or} \quad \frac{36}{100} \div \frac{4}{4} \quad \text{or} \quad \frac{9}{25} \quad \text{expressed as a simple fraction.}$$

? QUESTION

Show your work in figuring out the following problems.

1. $\frac{3}{4}$ = _____ (decimal fraction)

2. $\frac{9}{20}$ = _____ (decimal fraction)

3. 0.4 = _____ (simple fraction)

4. 0.15 = _____ (simple fraction)

5

In the previous discussion we saw that fractions are **parts** of the number **1**. Percents (%) are **parts** of the number **100**. Therefore, in order to convert a decimal fraction into a percent, we must first *multiply* the decimal fraction by 100.

The decimal fraction 0.23 is equal to 23%. (See the factors of 10—Multiplication Rule below if you need help with this.)

? QUESTION

Convert the following decimal fractions into percents, and show your work.

1. 0.79 = _____ %
2. 0.07 = _____ %
3. 0.18 = _____ %
4. 0.001 = _____ %
5. 0.809 = _____ %

FACTORS OF 10

Because the Metric System is based on units that differ from each other by factors of 10, we need to review how the decimal position *moves* when converting within metric units.

MULTIPLICATION RULE

When multiplying a number by 10, 100, 1000, etc., move the decimal position **to the right** by the number of 0's (zeros) in the multiplier. For example:

$$25 \times 10 = 25.0 = 250$$

By the way, you add 0's as you move the decimal point to an empty space.

$$25 \times 100 = 25.0\ 0 = 2500$$

$25 \times 1000 =$ _____

$2.5 \times 100 =$ _____

$0.25 \times 1000 =$ _____

$0.002507 \times 100{,}000 =$ _____

Try these for practice.

This rule allows you to convert **large** metric units like meters into **small** metric units like millimeters.

6

DIVISION RULE

When dividing a number by 10, 100, 1000, etc., move the decimal position **to the left** by the number of 0's (zeros) in the divisor. For example:

$$25 \div 10 = 2\,5. = 2.5$$

Again, you add 0's as you move the decimal point to an empty space.

$$25 \div 100 = 2\,5. = 0.25$$

$$25 \div 1000 = 0\,2\,5. = 0.025$$

See the Important Message below.

$$0.25 \div 10 = \underline{\hspace{2cm}}$$

$$2.5 \div 10 = \underline{\hspace{2cm}}$$

Try these for practice.

$$2.5 \div 1000 = \underline{\hspace{2cm}}$$

This rule allows you to convert **small** metric units like millimeters into **large** metric units like meters.

IMPORTANT MESSAGE

In science, the rule about decimal fractions like .025 is that you always put a "0" (zero) in front of the decimal point so that there is no confusion about the number being smaller than one. Therefore, we write the above example as 0.025.

METRIC PREFIXES

The Metric System is a standardized method for defining **length, volume,** and **weight**.

The standard metric units are **meter** (m), **liter** (l), and **gram** (g).

There are over a dozen prefixes used with each of these standard units that define "portions" of units and differ from each other by factors of 10. In this lab you will need to use only **three** of the metric prefixes. **Memorize them now!**

Prefix	Abbreviation	Compared to Standard	Examples
Milli	m	$\frac{1}{1000}$	mm, ml, mg
Centi	c	$\frac{1}{100}$	cm, *, *
Kilo	k	1000	km, *, kg

* cl, cg, and kl won't be used in this lab.

1. If you convert from a centi metric unit to a milli metric unit, are you going to a smaller unit or a larger unit in size?

2. How much difference is there between a *centi* unit and a *milli* unit?

3. How many decimal positions would you move for a conversion from *centi* to *milli*?

4. Converting from *centi* to *milli*, would you be moving the decimal position to the left or to the right? _____ You would be using the _____ rule.

5. Let's review your answers with an example.
   24.5 centimeters = _____ millimeters

6. Now, let's try it in reverse.
   53 millimeters = _____ centimeters

There will be more practice on these conversions as you go through the next sections on *standard* metric units.

LENGTH

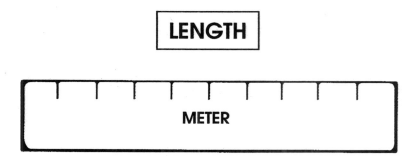

METER

The standard reference for **length** in the Metric System is the **meter** (abbreviated as m). You will be using three prefixes with this standard metric unit: milli ($\frac{1}{1000}$), centi ($\frac{1}{100}$), and kilo (1,000).

Remember to use the rules pertaining to the movement of the decimal point.

? QUESTION

1. The abbreviation for millimeter is **mm**. How many millimeters are in a meter? _____

2. The abbreviation for centimeter is **cm**. How many centimeters are in a meter? _____

3. The abbreviation for kilometer is **km**. How many kilometers are in a meter? _____
   (Aha! Did you get caught with your memory down because of the *wording* in this question? Think! Kilo means _____.)

4. How many meters are in a kilometer? _____

5. How many mm are in 1 cm? _____

6. How many cm are in a mm? _____

7. 40 cm = _____ m.

8. 40 cm is what fraction of a meter? _____

9. 40 cm is what % of a meter? _____

10. $\dfrac{32\ m}{10\ mm}$ = _____   **Hint:** You must have the same units on the top and bottom before doing the division.

11. 1.2 m x 30 = _____ m = _____ cm.

VOLUME

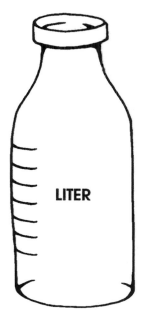

LITER

The standard value for **volume** in the Metric System is the **liter** (abbreviated as **l**).

The prefixes used for metric length units also apply to volume units. However, milli is the only prefix that we will use during this lab. The milliliter (**ml**) is a very common unit for the scientific measurement of small volumes of liquid.

? QUESTION

1. How many milliliters are in a liter? _____

2. How many liters are in a milliliter? _____

3. 355 ml = _____ l (This is a familiar volume for a canned soda or beer.)

4. 750 ml = _____ l (This is a familiar volume for wine bottles.)

5. 15 ml = _____ l (This volume is used in cooking—one tablespoon.)

6. What % of a liter is 15 ml? _____

7. $\frac{1}{2}$ liter = _____ ml

WEIGHT

The standard reference for *weight* in the Metric System is the *gram* (abbreviated as **g**).

The metric prefixes that we will use during this lab are milli and kilo.

GRAM

? QUESTION

1. How many grams are in a kilogram? _____

2. How many kilograms are in a gram? _____

3. How many milligrams are in a gram? _____

4. How many milligrams are in a kilogram? _____

5. 454 g = _____ kg (This is a familiar weight for a small loaf of bread.)

6. What % of a kilogram is 454 g? _____

7. 2.265 kg is the weight for a small bag of sugar. You are baking cupcakes for a school fund drive. It takes 100 g of sugar to make one batch of cupcakes. How many batches of cupcakes can you make with one small bag of sugar?

ACTIVITY #2

"CONVERSIONS: METRIC ⟷ ENGLISH"

You are familiar with the English System of measurement. In this Activity we will review some conversions between the English and Metric Systems.

LENGTH

GO GET

A combination meterstick/yardstick.

NOW

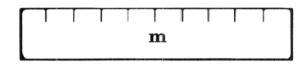

1. Make a line on a piece of paper exactly 10 inches long.

2. Measure that same line in centimeters. 10 inches = _____ cm.

? QUESTION

1. How many centimeters are in one inch? _____ cm = 1 inch

2. How many centimeters are in one foot? _____ cm = 1 foot

3. How many centimeters are in one yard? _____ cm = 1 yard
   (Check your answer with the measuring stick.)

VOLUME

GO GET

1. A 1-liter graduated cylinder.

2. A 10-milliliter graduated cylinder.

3. A 1-quart graduated cylinder.

4. An eyedropper.

5. A teaspoon.

1. Fill the container marked "1 quart" with water. (There may be a painted line indicating the exact 1 quart amount.)

2. Pour the 1 quart of water into a graduated cylinder for measuring liters.

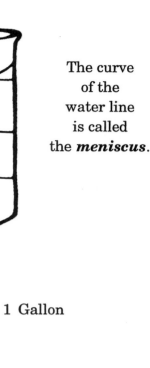

The curve of the water line is called the *meniscus*.

? QUESTION

1. How many ml are there in a quart? _____ ml = 1 quart

2. Which is the greater volume? (circle your choice)   4 Liters   or   1 Gallon

NOW

1. Count the number of drops of water it takes to fill the small graduated cylinder to the 1-milliliter mark.

$$1 \text{ ml} = \text{_____} \text{ drops}$$

2. Fill the 10-milliliter cylinder to the 5-milliliter mark. Pour that amount into the teaspoon.

$$5 \text{ ml} = \text{_____} \text{ teaspoon}$$

WEIGHT

? QUESTION

Try this weight problem. Home Depot bought sacks of cement from a company in Mexico. The sign above the cement display reads: "100 lb $4.95." The 100 lb cement sacks were also marked "45 kg."

1. How many sacks of cement can you buy for $4.95? _____

2. How many pounds are in one kilogram? _____

3. How much do you weigh in kilograms? _____

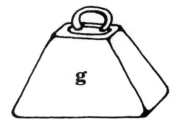

TEMPERATURE

The English measurement of temperature is in degrees Fahrenheit (°F). Using this scale, water freezes at 32°F and boils at 212°F.

There is a scientific temperature scale that is more like a metric scale. It uses Celsius (°C), and on this scale water freezes at 0°C and boils at 100°C.

NOW

This scale was copied from a dual scale thermometer. Using this illustration, answer the questions below.

? QUESTION

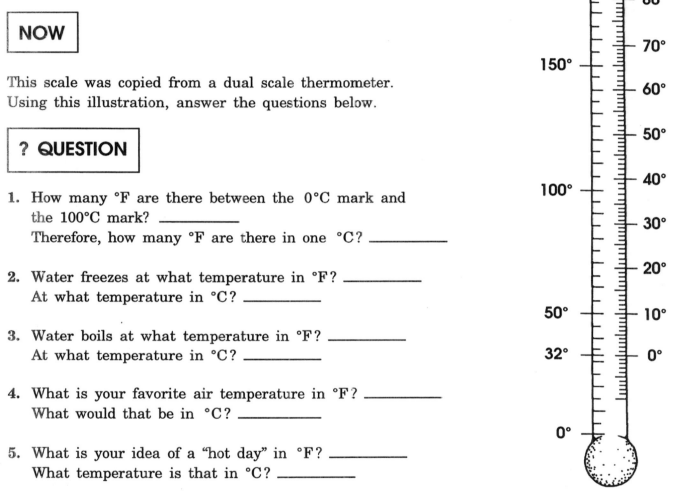

1. How many °F are there between the 0°C mark and the 100°C mark? _____
   Therefore, how many °F are there in one °C? _____

2. Water freezes at what temperature in °F? _____
   At what temperature in °C? _____

3. Water boils at what temperature in °F? _____
   At what temperature in °C? _____

4. What is your favorite air temperature in °F? _____
   What would that be in °C? _____

5. What is your idea of a "hot day" in °F? _____
   What temperature is that in °C? _____

6. Your normal body temperature is 98.6°F. Your child's forehead seems to be hot. You grab a Celsius thermometer by mistake and take her temperature. It reads 37°. Should you rush her to the hospital? _____ (The formula for converting °C into °F is: **°F = °C x 1.8 + 32**.) Why must the number 32 must be added?

7. Your cookbook says that roast beef is rare at 140°, medium at 160°, and well done at 170°. You like your beef cooked medium-rare and only have a meat thermometer in °C. What temperature will the thermometer have to reach for the roast to be done the way you like it?

8. The water temperature gauge on you new Volkswagen reads 85°C. Are you overheating your engine? _____ Explain.

CONVERSION FACTORS

(OPTIONAL EXERCISE)

Check with your instructor to see if you are responsible for doing this section before going on.

Conversion Factors (also called dimensional analysis in science) are a bit more complex than the methods used in the previous exercises. However, Conversion Factors are capable of solving both simple and complex problems in science, and they have an automatic self-checking feature when the rules are followed.

RULE 1   *The top unit of a Conversion Factor must be equal to the amount of the bottom unit of that factor.*

There are many Conversion Factors that you can use. Examples:

$$\frac{1 \text{ cm}}{10 \text{ mm}} \qquad \frac{10 \text{ mm}}{1 \text{ cm}} \qquad \frac{1 \text{ meter}}{39.4 \text{ inches}} \qquad \frac{1 \text{ liter}}{1000 \text{ ml}} \qquad \frac{100 \text{ pounds}}{45 \text{ kilograms}} \qquad \text{etc.}$$

Notice that you can invert any Conversion Factor.

All of the above factors are correct (OK factors) because the amount on the **top** of the factor is the *same amount* as that on the **bottom** of the factor.

$$1 \text{ cm} = 10 \text{ mm, so} \qquad \frac{1 \text{ cm}}{10 \text{ mm}} \quad = \quad \frac{\text{same}}{\text{same}} \quad = \quad \text{OK factor}$$

RULE 2   *The Conversion Factor that you choose must automatically cancel out the starting unit and leave the desired unit as your answer.*

For example, if you were starting with a kilometer unit and wanted to convert it into a yard unit, then the correct Conversion Factor would be:

$$\frac{1094 \text{ yards}}{1 \text{ kilometer}}$$

Remember: *Starting Unit   x   Conversion Factor   =   Desired Unit*

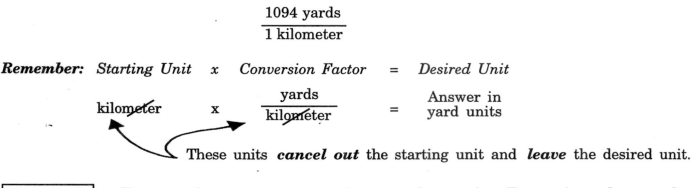

These units **cancel out** the starting unit and **leave** the desired unit.

RULE 3   *You may have to use more than one Conversion Factor in order to solve a particular conversion problem.*

Let's use the example above, and follow it through to see how Conversion Factors work. As you will see, knowing a little information and using a Conversion Factor can get you where you want to go.

In the example under Rule 2, we wanted to convert kilometers into yards. We decided that the necessary Conversion Factor must be:

$$\frac{yards}{kilometers}$$

But, what if you couldn't remember how many yards there are in a kilometer? The solution is: Use as many Conversion Factors as are necessary, starting with the one that you *do remember*.

Perhaps you remember that 1 meter = 39.4 inches. And you know that there are 1,000 meters in a kilometer. How can you use these Conversion Factors to answer the above question?

Start with what you know:

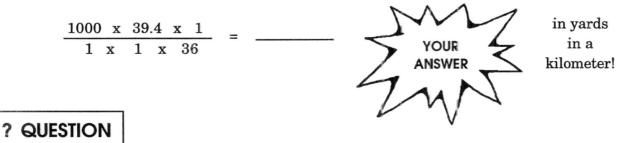

1 kilometer x $\frac{1000 \text{ meters}}{1 \text{ kilometer}}$ x $\frac{39.4 \text{ inches}}{1 \text{ meter}}$ x $\frac{1 \text{ yard}}{36 \text{ inches}}$ =

. . . and ***cancel*** the "units" as you work through the problem.

$$\frac{1000 \times 39.4 \times 1}{1 \times 1 \times 36} = \underline{\hspace{2cm}}$$

YOUR ANSWER

in yards in a kilometer!

? QUESTION

1. How many inches are there in 50 cm? ***Hint:*** There are 39.4 inches in a meter.

    What is the starting unit? _____

    What is the desired unit? _____

    What is your answer? (Show your Conversion Factors.)

2. How many liters are there in 10 gallons? ***Hint:*** There are 946 ml in 1 quart.

    What is the starting unit? _____

    What is the desired unit? _____

    What is your answer? (Show your Conversion Factors.)

ACTIVITY #3

"PERSONAL LIST OF METRIC REFERENCES"

Think of something easy to remember that you can associate with each of the metric units below.

Perhaps a centimeter might be the width of one of your fingernails.

Share your ideas among other members of your group. Be specific.

Whatever you choose as a reference, *make sure it's something you won't forget!*

MY METRIC LIST

Name: _____

| Metric Unit | My Personal Reference |

Length:

mm _____

cm _____

m _____

km _____

Volume:

ml _____ (How many drops?) _____

l _____

Weight:

mg _____

g _____

kg _____ (How many pounds?) _____

ACTIVITY #4

"A TEST OF YOUR MEASURING SKILLS"

HOW TO WEIGH AN OBJECT

Weighing balances are used to measure the weight of an object. It is important that the object to be weighed is put inside a weighing container so that the material is not spilled on the balance pan.

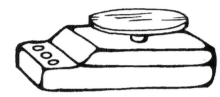
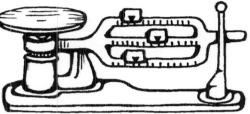

STEP 1 | "Zero" the scale. (Your instructor will show you how to do this.)

Then, weigh the weighing container.
Why is it important to weigh the container first?

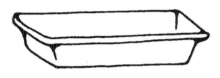

STEP 2 | Put the substance or object into the weighing container, and weigh them together.

STEP 3 | Determine the difference between the weights for Step 1 and Step 2.

GO GET

Some table salt.

NOW

Your instructor will give you specific directions for using each type of weighing scale. Make sure that you "zero" the scales before weighing.

1. Weigh 2.7 g of table salt, and put that amount onto a piece of paper. This is the recommended daily intake of salt in your diet.

2. Weigh 11.6 g of table salt, and put that amount next to the other pile. This is the typical daily salt intake by people in our society.

A TEST OF YOUR WEIGHING AND VOLUME SKILLS

It has been said that pennies before 1982 are heavier than pennies after 1982. We have 20 pennies in each category on the lab table.

EXPERIMENTAL QUESTION

1. Are the post-1982 pennies lighter or heavier?

2. If there is a difference in weight, then is that difference because either . . .
   a. they are not made of the same metal, or
   b. they are not the same size coin.

NOW

1. Design and perform an experiment to answer the questions above.

2. Compare groups of at least 10 coins in each category to measure the difference in weight or volume. **Hint:** The volume of objects can also be determined by measuring the displacement of water.

3. Ask to see one of the post-1982 pennies that has been cut in half to show its metal composition.

LAB REPORT

Question:

Hypothesis:

Experimental Design:

Results:

Conclusions:

MICROSCOPE

Summary Questions

1. Discuss the significance of the invention of the microscope.

2. How do you calculate the total magnification of an object being viewed through the microscope?

3. In a multi-layered object viewed through the microscope, how can you tell which layer is on top?

4. What are the diameters of the fields of view for the different magnifications of your microscope?

    low magnification at _____ X = _____ mm diameter

    medium magnification at _____ X = _____ mm diameter

    high magnification at _____ X = _____ mm diameter

5. Explain how to estimate the size of an object when viewed under the microscope.

6. Explain how you switch from one magnification to another magnification and still keep the same object in your field of view.

THE MICROSCOPE

INTRODUCTION

Physical reality is easiest to comprehend when we can see it. But, humans can't see very small or very far away things. We had to discover and explore the smaller world of atoms and the larger world of the universe by using indirect observations and our imaginations.

It wasn't until the inventions of the microscope and the telescope in the 1600's that humans were able to directly observe the worlds of the very small and the very large. Because of those inventions, biology and astronomy became major arenas of scientific exploration.

Your naked eye cannot see an object that is less than $\frac{1}{10}$ of a millimeter in length. But with a compound or dissecting microscope, small objects seem huge. The microscope is an extension of your visual sense. And this week, you will learn to use this tool and study the structure of organisms that ordinarily you cannot see.

ACTIVITIES

ACTIVITY #1

"FILM ON THE MICROSCOPE"

You will hear a short lecture or watch a film on the microscope before answering the questions below.

| GO GET |

A compound microscope. Unless you are extremely rich, *be sure to carry it with both hands!*

| NOW |

Work in groups of 3 or 4 to answer the questions below. Refer to Figures 1 through 6 on the following pages to help you in your discussions.

| ? QUESTION |

1. What happens to a beam of light when it passes through a lens? _____

2. If the light rays from an object enter the eye at a small angle, then the object will cover a _____ (smaller or larger) portion of the back of the eye.

3. The closer you move an object to your eye, the _____ (smaller or larger) the angle of light from that object entering your eye. Therefore, the object appears to be _____ (smaller or larger).

4. The magnifying lens _____ (increases or decreases) the angle of light coming from the object viewed. This results in _____ (increased or decreased) spread of the image at the back of the eye.

5. What is resolution?

6. Is resolution the same as magnification? _____

7. What is the total magnification if the objective lens is 4x and the ocular lens is 10x? _____

8. What happens to the size of the field of view as you change magnification?

9. Which objective lens gives the greatest magnification? _____

10. Which objective lens gives the greatest field of view? _____

11. When you search for a specimen, which lens should be used first? _____

12. How does depth of focus change with magnification?

FIGURE 1

CLOSE vs. FAR IMAGES

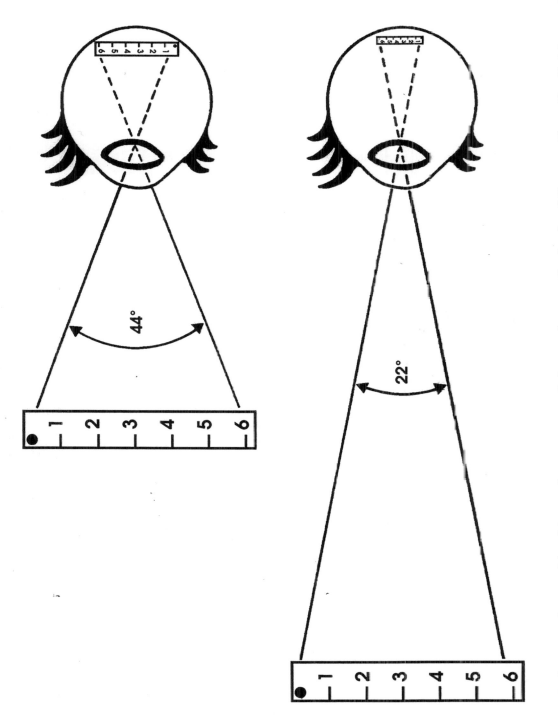

The light rays from a big object take up more of the surface at the back of the eye than the light rays from a small object. This is one way that our brains distinguish between small and large objects. Likewise, the light rays from a near object take up more of the surface at the back of the eye than the light rays from a distant object. This is why distant objects appear small compared to close objects.

It is the *angle* of light entering the eye that determines the amount of light spreading on the back of the eye. Big objects have a greater angle of spread than small objects. And near objects have a greater angle of spread than far objects.

FIGURE 2

CLOSE VS. MAGNIFIED IMAGE

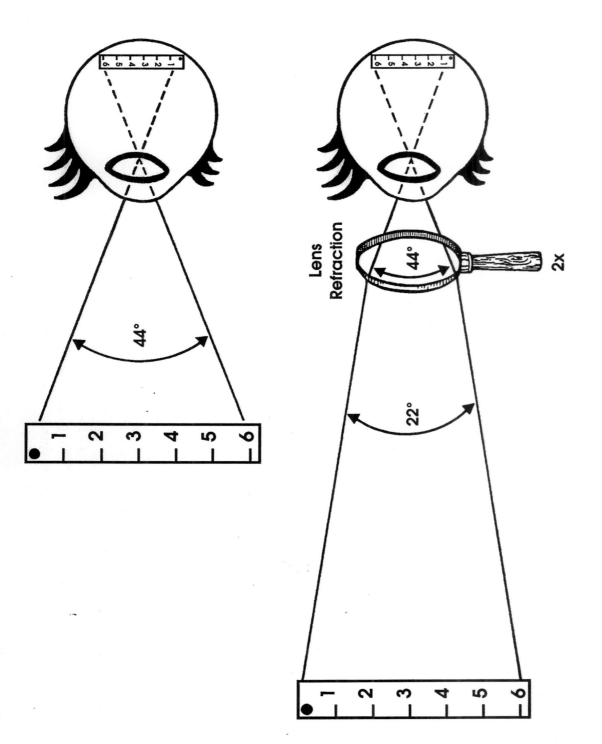

A magnifying lens bends the light rays traveling through it and *increases the angle* of light entering the eye, resulting in a bigger spread of the image at the back of the eye. Therefore, the object seems twice as close.

Because the image is spread over the entire back surface of the eye, more nerve cells are activated and much more *clarity of image* is sent to the brain. We see the object as bigger and we have excellent resolution at the same time.

FIGURE 3

RESOLUTION: SEEING IN FINE DETAIL

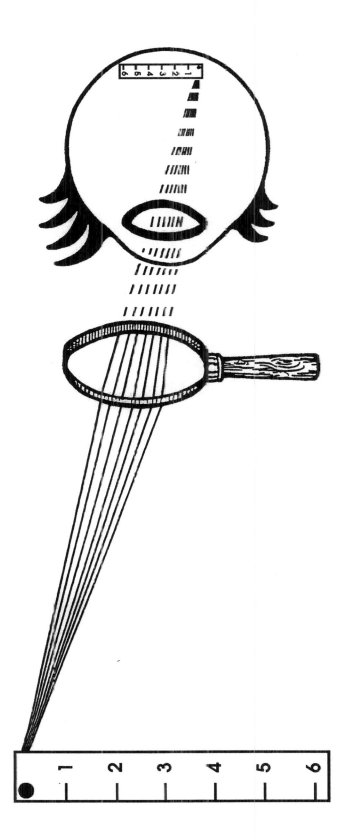

Resolution is the clarity of image produced by the microscope lenses. It is influenced by several factors. Shorter wavelengths of energy passing through the object result in better resolution. Electrons have a much shorter wavelength than visible light, which is why electron microscopes produce the best resolution. Also, a thin layer of oil between the object and the lens produces less random scattering (blurring) of light rays. Finally, excellent quality lenses refocus "stray" light rays from each point of the viewed object, which greatly improves resolution.

FIGURE 4

COMPOUNDED IMAGE

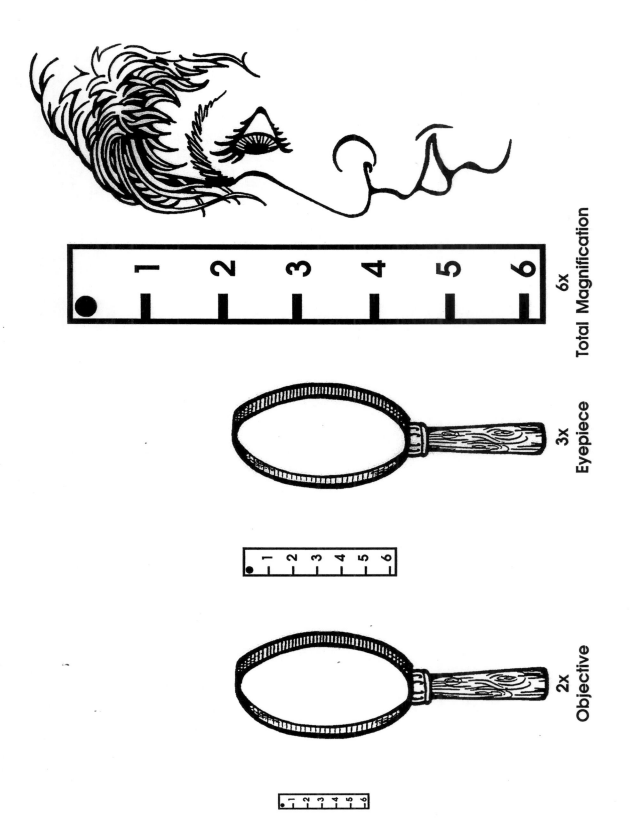

6x
Total Magnification

3x
Eyepiece

2x
Objective

When there are *two* lenses in a sequence, the image of the object is magnified *twice* before entering the eye. In this example, the first lens magnifies the object 2x, and the second lens magnifies that 2x image by 3x more. This results in a *compounded* magnification of 6x the original size.

FIGURE 5

MAGNIFICATION VS. FIELD OF VIEW

(Inverse Proportionality)

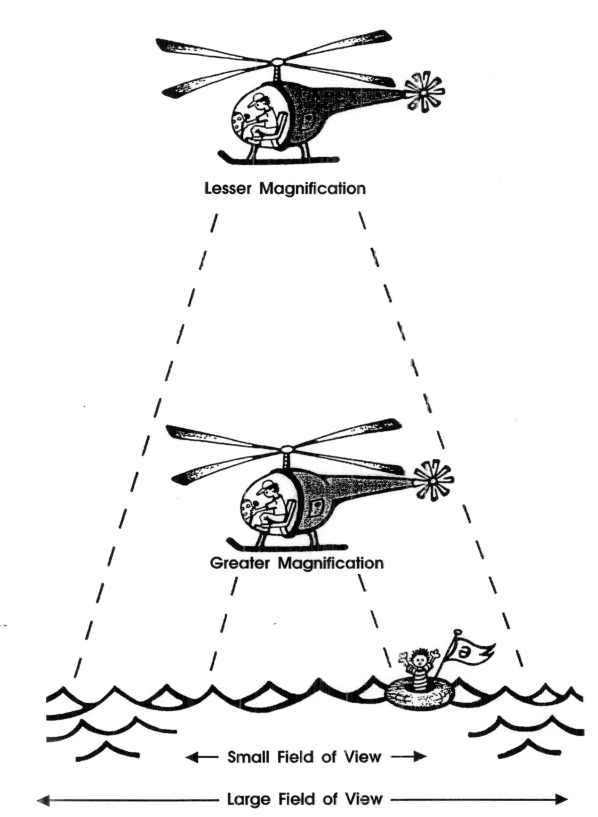

FIGURE 6

MAGNIFICATION vs. DEPTH OF FOCUS

(Inverse Proportionality)

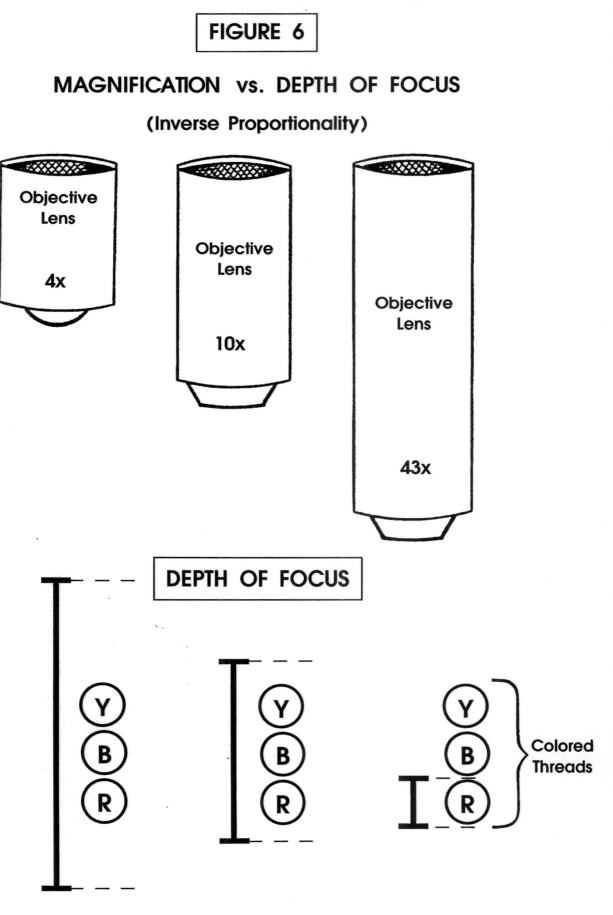

The *greater* the magnification the *less* the depth of focus.

Get familiar with the parts of a microscope by referring to the picture below and looking at your microscope. (Be able to name the parts and their functions on a Lab Test.)

Parts	Function
1. Nosepiece	A circular plate with 3 or 4 objective lenses that can be rotated into viewing position for different magnifications.
2. Objective Lens	This is the first lens that magnifies the image, and it is positioned just above the object being viewed.
3. Ocular Lens	This is the second lens that magnifies the image, and it is the one closest to your eye when you are looking through the microscope tube.
4. Coarse Focus Knob	Turning this knob decreases the distance between the objective lens and the stage, and allows you to focus the image entering your eye.
5. Fine Focus Knob	Use this knob to fine focus the image and when viewing through the higher magnifications.
6. Iris Diaphragm	This lever (or rotating disk) adjusts the amount of light shining through the object. Use just enough light to illuminate the object and give good contrast.
7. Stage	This is the platform designed to hold the microscope slide.

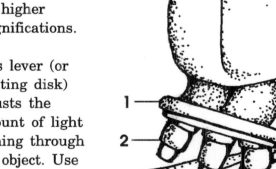

ACTIVITY #2

"A PINHOLE MICROSCOPE"

The pinhole microscope will demonstrate how the control of light rays determines your ability to see very small and close-up objects.

GO GET	

1. An 8" x 8" piece of black paper with a pinhole in the center.
2. A piece of paper with typed words on it.
3. A metric ruler.

NOW

In the first trial, don't use the pinhole paper. Hold the typewritten page in your left hand and slowly move it towards your eye. At some point the words on the page will become blurry. Have your lab partner measure the distance from the typed page to your eye. _____ cm.

? QUESTION

1. As you moved the printed page closer to your eye, did the *size* of the letters appear to change?

2. The words became blurry at some point close to your eye. What *quality* of your vision is being lost up close? _____ (We used this term to describe the clarity of the image.)

THEN

Repeat the experiment using the pinhole paper. Hold the pinhole paper with your right hand, and position the hole as close to your eye as possible. Move the typed page closer to your eye, and notice that the words which were blurry at _____ cm in the first trial are now readable at _____ cm this time.

? QUESTION

1. What *quality* of your vision has been improved by the pinhole paper?

2. Can you figure out how the pinhole microscope works? (Refer back to Figure 3 to help you with your answer.)

ACTIVITY #3

"FINDING AND FOCUSING"

GO GET

1. A compound microscope, one microscope slide, and one coverslip.

2. A cutout of the letter "e" from a piece of newspaper.

NOW

1. Turn on the microscope light.

2. Use the coarse focus knob to move the nosepiece all the way "up," and "click" the 4x objective lens into position. (Some microscopes raise the nosepiece, and others lower the stage. Check to see which microscope type you have.)

3. Make a slide with the letter "e". Place the letter on the glass slide, and slowly lower the coverslip over the letter to hold it in place. *Use no water.*

4. Place the slide on the stage, position the letter "e" over the stage hole, and secure the slide.

5. Look through the ocular lens and slowly turn the coarse focusing knob so that the distance between the lens and the slide decreases. The letter "e" will come into view.

6. Center the letter "e" in your field of view, and use the fine focus.

7. Adjust the iris diaphragm until you have the proper amount of light with good contrast. It is very important that you reduce the light so that the details of the object can be seen. *Using too much light is the most common mistake that students make when working with the microscope.*

? QUESTION

Normal View

1. Draw the letter "e" as it is positioned on the stage. Then draw the orientation of the letter "e" as it appears when viewed through the microscope.

2. What has happened to the orientation of the image as it passes through the lenses of the microscope?

Microscope View

3. When you move the slide forward on the stage, in what direction does the "e" appear to move when viewed through the microscope?

4. When you move the slide to the right, in what direction does the "e" appear to move?

5. What are your conclusions about the appearance and movement of an object when viewed through a microscope?

ACTIVITY #4

"DEPTH OF FOCUS: CROSSED THREADS"

Think back to the photographs that you have taken and looked at. There always seems to be a zone of sharp focus in the picture, with objects out of focus in front of and behind the sharp focus zone.

The sharp focus zone is called the **depth of focus**. This can become a serious problem when using a microscope as you increase the magnification. In fact, when looking through a microscope, the zone of sharp focus can be a major limitation in viewing the whole object. But, you can make use of that restriction as a tool to inspect *different layers* of the object being viewed.

Depth of focus is inversely proportional to magnification. In other words, as the magnification of the object increases, the depth of focus decreases. (Don't confuse this with *resolution,* which is a function of lens quality.)

GO GET

A prepared microscope slide of three crossed and colored threads.

NOW

Crossed
Threads

1. Review the directions in Activity #3 for finding and focusing with the 4x objective lens.

2. Once you have the crossed threads in focus—with proper contrast—be sure the point where the threads cross is *centered* in your field of view.

3. Turn the nosepiece until the 10x objective lens "clicks" into place. If you did a good job of centering the cross of the threads at low power, then the cross should be in your field of view at higher power. If not, go back to step #1 and try again.

4. Focus using the fine focusing knob, and adjust the contrast if necessary. Again, center the crossing threads in the field of view.

5. Turn the nosepiece until the high power (43x) objective lens "clicks" into place, and focus using the fine focusing knob. Adjust the contrast if necessary.

IMPORTANT MESSAGE

Do not make the mistake of raising the nosepiece before switching to the 43x objective lens!

If your threads were in focus at 10x, then the high power lens will *not* hit the slide.

6. Now, use the fine focusing knob to move the zone of sharp focus below the threads so that all the threads are blurred and slightly out of focus.

7. Then, reverse the direction you were turning the fine focusing knob. Watch the crossed threads until the *first* thread comes into focus. That will be the *bottom* thread.

    Hint: If you start out of focus *above* the threads, and move the lens closer to the slide, then the first thread to come into focus will be the top thread. However, if you start out of focus *below* the threads, and move the lens away from the slide, then the first thread to come into focus will be the bottom thread. Focus back and forth until you can determine the arrangement of the crossed threads.

Normally, focusing the highest power objective lens should be started slightly below the object (out of focus) and moved away from the stage so that you don't accidentally crunch the lens into the glass slide, cracking it and scratching the lens. After you become very practiced with a microscope, you won't make this mistake, but for now *please be very careful*.

? QUESTION

1. What color is the thread that is on the bottom, the middle, and the top?

    _____ top

    _____ middle

    _____ bottom

2. What microscope part is used to change the light intensity?

3. What happens to the light intensity as you change magnification of the objective lens?

4. When you switch to higher magnification, what should you do to the light intensity?

5. Describe the proper focusing technique in terms of moving the lens up and down when you first look at an object.

6. Why is this focusing technique so important?

ACTIVITY #5

"ESTIMATING THE SIZE OF AN OBJECT"

In order to estimate the size of an object, it is absolutely necessary to have an idea of the size of your field of view.

This applies to estimating the size of *any* object. For instance, if a person was standing next to a tree and you knew the size of the tree, then you would have a standard against which you could estimate the size of the person.

The person is about _____ (what fraction) the size of the tree. Therefore, the person is about _____ meters tall.

The standard we use to estimate the size of an object when viewed under the microscope is the *diameter* of the *field of view*.

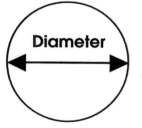

The field of view looks like this.

And the *diameter* of the field of view is the length of a straight line through the center of the circle.

 GO GET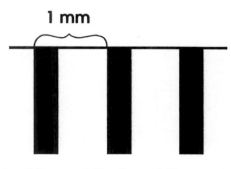

A transparent 15-centimeter ruler.

NOW

1. Each dash on the metric side of this ruler is a millimeter unit. Under the microscope these dashes appear wide. To properly measure the diameter of the field of view under a microscope, you must measure a millimeter as the distance from one side of a mark to the same side of the next mark.

1 mm

Ruler Viewed Under a Microscope

34

2. Using the ruler, measure the diameter of the field of view with the 4x objective lens in position. Estimate this diameter to the nearest $\frac{1}{10}$ of a millimeter.

Diameter of 4x field of view = _____ mm.

3. Since the diameter of the field of view is *inversely proportional* to the magnification, we can use this knowledge to estimate the diameter for the other two higher power lenses.

Remember: As magnification *increases*, the field of view *decreases*.

? QUESTION

1. If the diameter of the field of view through the 4x lens is 4mm, then under 10x it will be

Larger than 4mm   or   Smaller than 4mm

2. What is the diameter of the field of view you measured through the 4x lens?

_____ mm

3. Instead of physically measuring the diameter of the 10x view, you can use a calculation to give you the answer. Because the 10x objective lens provides $2\frac{1}{2}$ times *more* magnification than the 4x objective lens, its field of view is $2\frac{1}{2}$ times *smaller* ($10 \div 4 = 2\frac{1}{2}$).

So, use the above calculation to figure out how big the diameter of the field of view would be through the 10x lens. *Remember:* The field will be $2\frac{1}{2}$ times smaller than at 4x.

_____ mm

Note: If your microscope does not have a 43x objective, substitute the magnification of the lens that you do have for all of the calculations below where the number "43x" occurs.

4. This calculation method applies to the 43x objective lens as well. 43x is _____ times *more* magnification than the 4x objective. So, the diameter of the field of view at 43x will be _____ times *smaller* than at 4x.

5. What is the diameter of the field of view for your microscope through the 43x lens?

_____ mm

GO GET

1. A glass slide and a coverslip.

2. The small letter "i" cut from a newspaper. Be sure to include the dot.

1. After making a slide of the "**i**" like you did in Activity #3, look at it under low power and center the dot in your field of view. Switch to the 10x objective lens, center the dot, then go to the 43x lens. Estimate how much of the field of view is occupied by the dot through the 43x lens.

_____ %

2. If the dot is about _____ % of the 43x diameter of view, then the diameter of the dot is approximately _____ mm.

ACTIVITY #6

"SEARCHING THE FLY WING"

This Activity will test your ability to look for something specific under the microscope.

| GO GET |

A prepared microscope slide of a fly wing.

| NOW |

1. Look at the picture of the fly wing at the front of the lab. Study it carefully. A particular vein intersection will be marked on the picture.

2. If it helps you to remember, draw a simple sketch from the picture.

   Be sure to indicate the field of view as it appears in the picture.

3. Find the *exact* vein intersection in your microscope slide *under the same magnification* as you see in the picture.

4. Be able to do this exercise on a Lab Test.

Fly Wing Veins

Summary Questions

1. Discuss the Cell Law and explain why it's discovery was so important.

2. How big is a human cheek cell compared to the dot over the letter "i"? (State your answer as a percent of the diameter.)

3. How do you know whether a cell is eukaryotic?

4. How long in mm is a typical onion cell?

5. How many layers of cells are in the Elodea leaf?

6. Which layer of the Elodea leaf has thick-walled cells?

7. Where are color pigments located in the cells of plants? Illustrate your answers.

8. What are the stomata, where are they found, and how are they important?

CELLS: A RADICAL IDEA

INTRODUCTION

Educated people of the early 1800's had the belief that some invisible force was responsible for life in all its growing and changing forms. In 1839, Schleiden and Schwann presented the idea that *cells* were the creative force responsible for living organisms. Even though it was a radical change from the beliefs of the time, their ideas were accepted almost without question. The logic and proofs presented made perfect sense. Many puzzling questions were now answered.

The **Cell Law** states:

1. all of life consists of cells,
2. all cells come from previous cells, and
3. all life processes derive from cellular activities.

The implications of this principle are profound. It means that:

1. all life forms are related to each other at the cellular level,
2. all functions of organisms (including humans) are based on individual cell activities, and
3. all cellular activities are based on chemical processes.

Yet, in spite of what it implied, scientists and laymen were in agreement about cells. How different was the reception of Schlieden and Schwann when compared to other pioneers in biology, like Gregor Mendel in genetics research, and Charles Darwin in concepts of evolution.

This week you will look at cells and discover some of their general features.

ACTIVITIES

ACTIVITY #1

"HOW TO MAKE A WET MOUNT SLIDE"

In order to observe cells, you will have to become good at the technique of making a slide. This requires patience and careful handling of equipment. Take your time.

STEP 1 You will need a microscope slide and a coverslip.

STEP 2 Put a *drop* of water on the slide.

STEP 3 Put the object into the drop of water. The object must be *very* thin. You will see the importance of this when you make a wet mount of onion cells.

STEP 4 Place the coverslip over the object by first placing one edge down, and then slowly lowering the other side so that you don't trap air bubbles. Air bubbles will look like discarded tires, and are actually quite interesting in appearance, but they will interfere with your view of the object you really want to see.

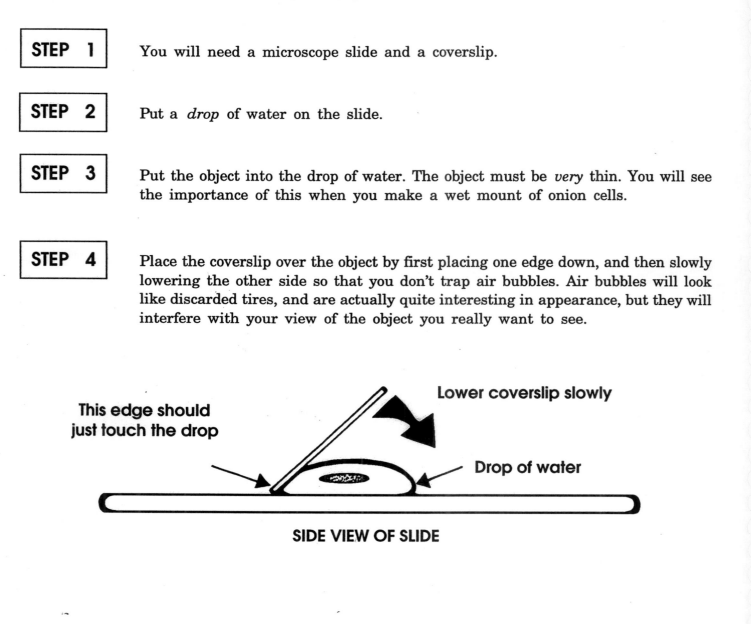

This edge should just touch the drop

Lower coverslip slowly

Drop of water

SIDE VIEW OF SLIDE

IMPORTANT MESSAGE

Whenever you make a slide of something during this semester, you should use the wet mount method. It is the very best way to get a clear view of the object, and it prevents the specimen from drying out.

40

ACTIVITY #2

"HUMAN CHEEK CELLS"

There are two types of cells: *prokaryotic* and *eukaryotic*. All of the cells you look at in the lab today are of the type called **eukaryotic** (true nucleus). The eukaryotic type of cell is the basic component of all multi-celled life forms. Some eukaryotic cells, such as the *Paramecium*, are single-celled organisms.

However, the largest number of single-celled organisms are the bacteria and their relatives, and they are of another cell type called **prokaryotic** (before nucleus). You will have the opportunity to investigate them later in the semester.

Now you will begin your journey into the world of cells by looking at *human cheek cells*.

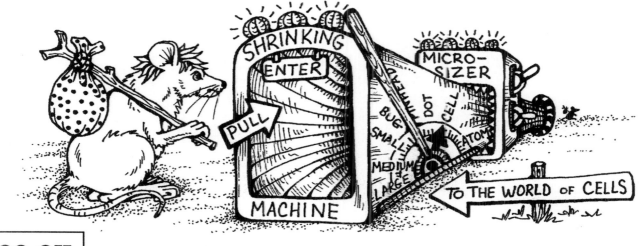

GO GET

1. A compound microscope.

2. A microscope slide and coverslip.

3. A toothpick.

NOW

1. Put a small drop of water on the microscope slide.

2. *Gently* scrape the inside of your cheek with the blunt end of the toothpick. You will have collected hundreds of eukaryotic cells on the toothpick.

3. Pay attention to exactly where you put the cells on the slide. They are hard to find. *Throw away the toothpick into the special waste container or disinfectant solution.*

4. Cover the drop with a coverslip.

5. Look at the cells under high power with your compound microscope. **Remember:** They will be very hard to see.

6. Now, put a drop of *methylene blue* stain at the edge of the coverslip. Use a small piece of tissue paper at the other edge of the coverslip to absorb the excess fluid and pull the stain across the slide.

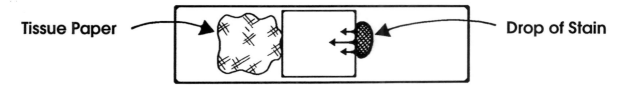

Tissue Paper   Drop of Stain

This method allows you to apply a stain without removing the coverslip.

7. Look at the cells again under high power. You should be able to see the **nucleus** and the **cell membrane**. The nucleus controls the cell functions, and the cell membrane controls what molecules go into and out of the cell.

8. The advantage of stains is that we can see structures better. The disadvantage is that stains kill the cells. *Never use a stain if you want to see living cells!*

9. Draw a simple sketch of your cheek cells. Label the cell membrane and the nucleus.

Human Cheek Cells

? QUESTION

1. How do you know that your cheek cells are *eukaryotic* cells?

2. You may be asked on a Lab Test to make a wet mount of cheek cells, and find them under the microscope. Can you do it?

3. Point out the cell membrane and cell nucleus to your lab partner. Be able to do the same for your instructor on a test.

FINALLY

Put the slide and coverslip into the disinfectant solution. You won't reuse these slides.

ACTIVITY #3

"ONION CELLS"

In this Activity, you will be examining a plant cell. As you work through the steps, notice a difference between plant and animal cells (human cheek cells).

GO GET	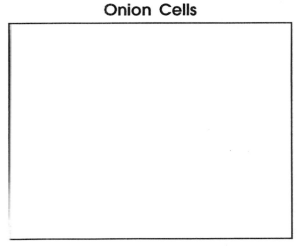

1. Cutting board and knife.

2. Cut an onion into "onion rings."

NOW

1. On the inside of each onion piece you should be able to peel off a one-cell-thick layer of tissue. It will look like a piece of plastic wrap. You can use a razor blade or forceps to start the peel. But *don't slice* off a piece; that will be many cell layers thick. You need a one-cell layer.

2. Place the onion peel into a drop of water on the slide, trying not to fold it over on itself.

3. Finish the wet mount, and look at the cells under the compound microscope (low power first, then high power).

4. Now, put a drop of *iodine* stain at the edge of the coverslip. Use the tissue paper to draw the stain across.

5. Look at the cells again. You should be able to see the **nucleus** and the **cell wall**. The nucleus controls cell functions, and the cell wall is a little box made of cellulose (wood) produced by the cell for support. *Draw a simple sketch of the onion cells at high power. Label the cell wall and the nucleus.*

Onion Cells

? QUESTION

What differences and similarities did you observe between the onion (plant) cells and the cheek (animal) cells?

FINALLY

Wash off the slide and coverslip so that you can use them for the next Activity. **Remember:** Slides and coverslips cost money. Don't throw them away. They can be used over and over again.

ACTIVITY #4

"PARAMECIUM"

A *Paramecium* is a one-celled organism. Because it must do everything in its life as only one cell, it is far more complex than any single human cell.

GO GET

A drop of water from the bottom of the *Paramecium* culture (these organisms usually settle near the bottom), and put the drop on your slide.

NOW

1. Make a wet mount, and find the *Paramecium* under the 10x objective lens (that will be 100x magnification).

2. Your job is to train your hands to be able to follow this organism under the microscope. **Hint:** Don't think! Let your hands work by themselves.

3. Practice your quick-vision skills by making a sketch of the *Paramecium*. Label the nucleus and the cell membrane.

Paramecium

4. **Optional:** We have a product called Protoslo® that can be added to the water drop sample. It will dramatically slow the movement of the *Paramecium* because it thickens the water. However, Protoslo® will also push the one-celled organisms to the outside perimeter of the water drop. You have to first mix the Protoslo® with the drop of sample.

? QUESTION

1. What obvious differences did you observe in the one-celled organism as compared to one cell of a multi-celled organism (cheek or onion cells)?

2. If you were asked on a test to find a *Paramecium* in "pond water," could you do it? **Remember:** Pond water will contain many different organisms.

3. How do you think a *Paramecium* eats? (Refer to your textbook for the answer.)

FINALLY

Wash off the slide and coverslip in preparation for the next Activity.

ACTIVITY #5

"ELODEA LEAF"

Elodea is found in freshwater ponds, and is commercially grown and sold as an aquarium plant. Pay attention to the differences between the *Elodea* cells and the onion cells that you observed in Activity #3.

GO GET	

An *Elodea* leaf from near the tip of a healthy plant. Use forceps to pluck the leaf. Keep track of which side is the *upper* side of the leaf.

NOW

1. Make a wet mount with the upper side of the leaf facing up, and find the *Elodea* cells under the 10x objective. Look near the tip of the leaf. These are often very active cells. Then switch to high power magnification.

2. You will know that the cells are alive and active if you can see **chloroplasts** moving around the cell. It may take a few minutes of warming under the microscope light before the chloroplasts begin to move. Chloroplasts are the cell structures (organelles) that do *photosynthesis* (food production) in plants.

Elodea Cells

3. There is a large sac of fluid inside the *Elodea* cell called the **central vacuole**. Imagine a swimming pool that has a huge clear sac of water floating in it. You can't actually see the sac of water, but the movements of everyone in the pool will be influenced as they bump into that large clear sac.

   Now watch the movement of chloroplasts. See if you can observe the indirect evidence that the central vacuole is in the cell and is influencing the movement of those chloroplasts.

   Draw a sketch of the *Elodea* leaf cell and show the chloroplasts.

4. During the Microscope Lab you used the fine focus at high power to determine which color of thread was on top of the others.

   An *Elodea* leaf is two cells thick. You should be able to decide whether the top layer is made of bigger or smaller size cells than the bottom layer. Do this now.

1. What color are the chloroplasts?

2. When you see the color of plants, what *structures* are you actually seeing?

3. What does the *movement* of the chloroplasts tell you about the cell?

4. If you can't actually see the central vacuole inside the *Elodea* cell, then how do you know that it is there?

5. *Elodea* leaves are two cells thick. One of the layers has thick cell walls. Those cells are in the . . . (circle your choice)

    Top layer   or   Bottom layer

6. What is the most obvious difference you observed between the *Elodea* cell and the onion cell?

    What does that observation tell you about the activity of food production in the onion cell?

    Where is food produced in the onion plant?

FINALLY

Don't wash off your slide! Save this *Elodea* slide for the next Activity. You will want to look at the cells one more time.

ACTIVITY #6

"COMPARISON OF COLOR IN FLOWER PETAL, RED-ONION SKIN, AND *ELODEA* LEAF"

The color of plant parts is determined by various *pigments* inside the cells. In some cells the color is inside organelles called **plastids**. Chloroplasts are one kind of plastid. There are other plastids that contain different colored pigments.

In other cells the pigment is distributed throughout the *water* of the central vacuole. Your task in this Activity, is to determine and compare where the color is located in an *Elodea* leaf, red-onion skin, and yellow flower petal.

In order to determine whether a pigment is in the central vacuole or inside the plastids, you must look at one cell layer, and note the distribution of the color within the individual cells.

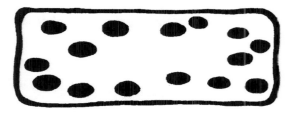

A color distribution like this indicates that the pigment is inside the plastids.

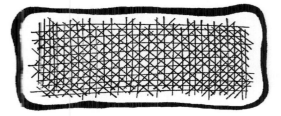

A color distribution like this indicates that the pigment is in the water of the central vacuole.

GO GET

1. A small piece of the outside *red* skin of the onion.

2. A yellow flower petal.

3. Your *Elodea* leaf slide from Activity #5.

4. Two more slides and coverslips.

NOW

1. Make a wet mount of a *one-cell-thick* layer of the red-onion skin.

2. Determine whether the red color is inside the plastids or distributed throughout the water of the central vacuole.

THEN

1. Make a tearing peel of the yellow flower petal.

   The ragged edge will be one cell thick.

2. Make a wet mount of the yellow flower petal, and look at the *one-cell-thick* area on the ragged edge of the tearing peel.

3. Determine whether the yellow color is inside of the plastids, or distributed throughout the water of the central vacuole.

? QUESTION

1. Look again at your slide of the *Elodea* leaf. Where is the green color in *Elodea* cells?

2. Where is the red color in the red-onion skin cells?

3. Where is the yellow color in the yellow flower petal cells?

4. If you were asked on a Lab Test to determine where the color is in red rose petal cells, or orange flower petals, could you do it and show your evidence (including your skill at making a *one-cell-thick* wet mount)?

FINALLY

Wash off your slides and coverslips. Save one for the next Activity.

ACTIVITY #7

"*ZEBRINA* LEAF EPIDERMIS WITH STOMATA"

Most animals have some method of breathing. Do plants have any such equivalent process?

Scattered throughout the *underside* skin of the *Zebrina* leaf are small openings called **stomata**. Stoma is the Greek word for "mouth." These openings look like green lips.

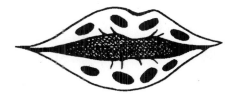

The stomata regulate the flow of air into and out of the leaf. Your job is to find these stomata.

GO GET

A *Zebrina* leaf.

NOW

1. Make a leaf peel with the leaf *upside down* so that you can get a *one-cell-thick* peel of the **underside** of the *Zebrina* leaf. A thin layer will peel off the bottom of the leaf as you tear if you are doing the procedure correctly. Cut off the thin layer piece with a razor blade. Don't try to make a thin slice with your razor blade. You won't get a single-layer slide. Always use the tear/peel technique.

2. Make a wet mount and look for the stomata.

3. Look closely at the structure of the stomata. Notice whether you can identify the organelles inside of the two cells that make up the stomata. These two stomata cells are called *guard cells* because they "guard" the opening. (Your textbook discusses the details of the chemical processes by which the stomata are opened and closed.)

4. Look at the skin cells around the stomata. Notice what cell organelles they *don't* have.

5. Draw a simple sketch of the *Zebrina* leaf stomata and the surrounding cells.

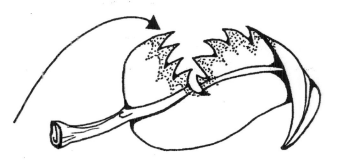

Stomata

1. *Zebrina* leaf stomata perform a specific function. What is it?

2. What organelles do the guard cells contain that are absent in the skin cells of the leaf?

3. Why would the guard cells have chloroplasts when the other skin cells of the leaf don't have chloroplasts?

4. If you were asked on a Lab Test to make a *one-cell-thick* wet mount of leaf stomata, could you do it?

FINALLY

Wash off your slide and coverslip, and return them to the supply table. Return your compound microscope to the cabinet.

Summary Questions

1. Why is an understanding of chemistry important?

2. Discuss the difference between element, atom, and molecule.

3. The Atomic Numer is the number of _____ , in an atom.

4. The Atomic Mass is the number of _____ , in an atom.

5. The number of protons in an element also equals the number of _____ .

6. Isotopes have different numbers of _____ . Why are isotopes important?

7. Ions have different numbers of _____ . Why are ions important?

8. How many atoms of Group I would react with one atom of Group V?

9. How many atoms of Boron (B) would react with how many atoms of Sulfur (S)? (*Hint:* check their group numbers.)

CHEMISTRY CONCEPTS

INTRODUCTION

The first principles of the universe are atoms and empty space;
everything else is merely thought to exist.

—Democritus (~ 400 B.C.)

Chemistry is the science that deals with the nature of and the changes in the composition of matter. The relationship of particles within the atom, and the interactions between the atoms account for everything that we call "matter" in our world. At first it seems impossible to believe that air, liquid, and solid could be made of the same basic particles. But when we watch a cube of ice melt in a pool of water, then drip onto a hot plate and become vapor, we are forced to conclude that there is something going on that we can't explain in any other way.

Our eyes can't see the particles but we know they must be there—in another world—in a world of atoms.

This lab is meant to be an aid to your understanding of some of the chemical processes and interactions that will be discussed throughout your study of biology. Here we will start the discussion. And we hope this lab will encourage you to continue your exploration into the science of chemistry.

ACTIVITIES

ACTIVITY #1

"THE PERIODIC TABLE OF ELEMENTS"

The word **element** originally came from a Latin word meaning "first principle." In Roman times, people thought that the universe was made up of four basic elements: earth, air, fire, and water.

Today, the modern chemist defines an *element* as the most basic kind of substance which cannot be broken into simpler substances by ordinary chemical processes. Over the years, more than 100 elements have been discovered and described.

Water is the most abundant material on the surface of the earth. However, using our definition, we don't call water an element because it can be broken down into two simpler substances: *hydrogen* and *oxygen*. Hydrogen and oxygen cannot be broken down into simpler substances, so we conclude that they are elements.

OBSERVE

Study the "Abbreviated" Periodic Table of Elements on the next page for a few minutes. Notice how the different elements are listed, grouped, and numbered.

All living things are chemical combinations based on the lighter elements. These are shown in this "Abbreviated" Periodic Table. (A complete Table of Elements can be found in your textbook or the lab room.)

During the formation of the earth, heavy elements sank into the inner molten core and were not available for use in chemical processes taking place on the surface of the planet. Look again at the Table of Elements. These are the common elements that were available in the "Soup of Life" from which you came.

? QUESTION

1. Do you know what the numbers on the periodic chart mean?

2. Do you know why the elements are listed in certain groups?

3. Do you recognize the names of any elements on the chart by their alphabetical symbols?

4. If you have access to a complete list of element names (in your textbook or on a lab chart), then you will notice that the element symbol is not necessarily an abbreviation of the word we use for that element. Can you find the common name for the element symbols Na and K?

"ABBREVIATED" PERIODIC TABLE OF ELEMENTS

Group I	Group II	Group III	Group IV	Group V	Group VI	Group VII	Inert Gases
1 **H** 1							2 **He** 4
3 **Li** 7	4 **Be** 9	5 **B** 11	6 **C** 12	7 **N** 14	8 **O** 16	9 **F** 19	10 **Ne** 20
11 **Na** 23	12 **Mg** 24	13 **Al** 27	14 **Si** 28	15 **P** 31	16 **S** 32	17 **Cl** 35	18 **Ar** 40
19 **K** 39	20 **Ca** 40						

ACTIVITY #2

"ATOMS AND ISOTOPES"

ATOMIC STRUCTURE

One of the most important questions in chemistry is: What makes one element different from another element?

The smallest particle of an element that can exist and still retain the chemical properties of that element is called an **atom**. Today we know that there are particles even smaller than atoms called **protons, neutrons,** and **electrons**. It is the various combinations of these subatomic particles that control the chemical properties of the different elements.

All atoms consist of an inner core that contains most of the mass of the atom, and an outer zone in which there are very light particles, called electrons, that are either like energy waves or like particles, depending on how they are studied. If you monitor electrons with a machine that looks for particles, they will appear to be particles. And if you try to find them as energy waves, that machine will record them as energy waves. Their nature is both particle and wave.

The inner core is called a **nucleus**, and it is made up of two kinds of particles: *protons* and *neutrons*. The outer zone of the atom is filled with those tricky, high energy particles called *electrons*.

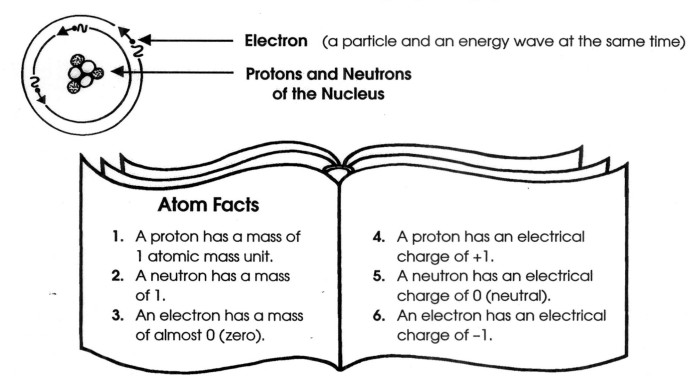

Electron (a particle and an energy wave at the same time)

Protons and Neutrons of the Nucleus

Atom Facts

1. A proton has a mass of 1 atomic mass unit.
2. A neutron has a mass of 1.
3. An electron has a mass of almost 0 (zero).

4. A proton has an electrical charge of +1.
5. A neutron has an electrical charge of 0 (neutral).
6. An electron has an electrical charge of –1.

In this lab we must simplify definitions and explanations. A chemistry class would look at the concepts of weight and mass with important differences. Here, we consider them as equals. (This is, in fact, not true.)

ATOMIC NUMBER AND ATOMIC MASS

The Periodic Table of Elements contains the atomic mass and atomic number of each element. These numbers indicate the number of protons, neutrons, and electrons in each element's basic structure.

OBSERVE

1. Each square in the Periodic Table represents an *element*.

2. There are two numbers in each square.

   a. The smaller number at the top of the box is the **atomic number**.

   b. The number at the bottom of the box is the average of all isotopes of the element. If you "round off" this value to the nearest whole number, then that number is the **atomic mass**. On your "Abbreviated" Periodic Table we have provided the rounded off atomic masses.

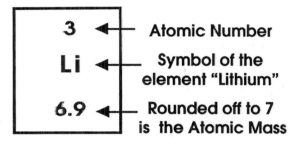

3 ◄—— Atomic Number

Li ◄—— Symbol of the element "Lithium"

6.9 ◄—— Rounded off to 7 is the Atomic Mass

INFORMATION

Atomic Number = # of Protons

Atomic Mass = # of Protons + # of Neutrons

A single atom has an overall neutral charge because the number of + charges (protons) is equal to the number of − charges (electrons) in the atom.

Hint: If you know the atomic number, then you automatically know how many protons or electrons are in that atom. This is because the number of protons is equal to the number of electrons.

? QUESTION

1. Using the "Abbreviated" Periodic Table of Elements, put the appropriate symbol by the element's name, and determine the number of protons, electrons, and neutrons.

Element	Symbol	# of Protons	# of Electrons	# of Neutrons
Lithium				
Beryllium				
Boron				
Carbon				
Nitrogen				
Oxygen				
Fluorine				

2. Use the chart you just completed to answer the following questions.

    a. What is the atomic number for carbon? _____

    b. What is the atomic mass for lithium? _____

    c. What is the atomic number for nitrogen? _____

    d. What is the atomic mass for oxygen? _____

    e. Fe is the symbol for the element ferrous, commonly known as iron. Knowing that information, what is the name for the chemical structure pictured here?

        Answer: a ferrous wheel.

ISOTOPES

Elements exist in different forms called *isotopes*. You may have heard of radioactive iodine which is used in medicine, heavy water which is used in nuclear reactors, or carbon 14 which is used in paleontology to date fossils. All of these are isotopes of the elements iodine, hydrogen, and carbon, respectively.

Isotopes of an element usually have the same chemical properties. Isotopes have the *same* number of *protons,* but have *different* numbers of *neutrons.* That means that they will have slightly different *atomic weights.*

The atomic weight as recorded in the Periodic Table is the average weight of all the various isotopes of that element. That value is recorded at the bottom of each element box.

The mass (protons + neutrons) of individual isotopes is not included in the Periodic Table, but instead is indicated by putting a number to the upper left of the element's symbol. For example, ^{12}C is an isotope of carbon and is called carbon twelve because its atomic mass is twelve. ^{14}C has an atomic mass of 14; it is called carbon fourteen.

Because of its radioactive properties, ^{14}C can be put into a sugar molecule, and that molecule can be "followed" through your metabolism to determine how you process your food Calories.

$^{14}C_6H_{12}O_6$
SUGAR

Isotopes are often used as a chemical "tag" to follow the molecules of a particular substance through a biochemical process. For example, in medicine, radioactive iodine is injected into the the bloodstream and the isotope is tracked with special machines as it filters into the patient's kidney and urine. It is used to discover kidney disease or kidney malfunction.

1. How many neutrons are in ^{14}C ? _____ Its atomic mass is _____.

2. How many neutrons are in ^{12}C ? _____ Its atomic mass is _____.

3. ^2H is called deuterium. It is the isotope of hydrogen that is used to make heavy water for nuclear power plants.

    How many neutrons are in ^2H ? _____

    Its atomic mass is _____.

4. Why do the isotopes of an element have the same basic chemical properties?

5. What is the primary use of isotopes in medicine and biological research?

ACTIVITY #3

"MOLECULAR FORMATION"

THE EIGHT GROUPS OF ELEMENTS

Look at your "Abbreviated" Periodic Table of Elements. About 150 years ago, chemists discovered that all of the known elements could be arranged from low atomic weight to high atomic weight. And, when they were put in this order, it was noticed that there were repeating physical and chemical properties. It was amazing! There was some kind of "order" in the universe that could be seen through the study of chemistry.

So, the Periodic Table of Elements was constructed on those repeating patterns. The elements in Groups I, II, and III are all *metals* (except for hydrogen, which chemically reacts like a metal but is a gas). The other groups are called the *non-metals*.

Metals easily *release* electrons to other groups. The *group number* of the metals indicates *how many electrons* can be released: 1, 2, or 3.

The non-metal elements in Groups IV through VII *attract* or *share* electrons from other elements. Now, if you take the group # of a non-metal element and subtract it from the number *eight*, then you will know the *number of electrons* that this element will attract or share.

For example, the element oxygen will attract two electrons when it combines with other elements to form molecules.

$$(8 - Group\ VI = 2)$$

To complete our discussion of the groups in the Table, the *inert gases*—the final group—get their name because they *do not* react with any other element.

The key to understanding *chemical bonding* comes from understanding the interactions between the electrons of different elements when they bump into each other. Do they create a chemical reaction, or is there no chemical reaction? The number "8" turns out to be a very important clue in predicting molecular formations in chemistry.

NOW

Keeping in mind what you have just learned about the Periodic Table of Elements, answer the following questions.

? QUESTION

1. Does sodium (Na) attract or release electrons? _____

   How many electrons are involved in the transfer? _____

2. Does chlorine (Cl) attract or release electrons? _____

How many electrons are involved in the transfer? _____

3. What do you think would happen if a sodium atom and a chlorine atom bumped into each other?

4. Welders use helium (He) and argon (Ar) gases to blow over the metal during the welding process. Why would they want to use these gases?

5. If 10 magnesium (Mg) atoms and 10 chlorine (Cl) atoms were allowed to bump into each other, could they combine to form a substance? _____

How many molecules would be made? _____

Would there be any atoms left over? _____

Which ones? _____ How many? _____

6. My friend, the inventor, says that she has just made a translucent, light-weight metal by combining aluminum with helium and silicon. She wants me to invest $10,000 in the process, and promises that we will be millionaires in only six months. What do you think I should do? Why?

IONS

An atom has the same number of positive charges (protons) as negative charges (electrons). Some atoms can lose electrons (Groups I, II, and III), and other atoms can gain electrons (Group VII). An *ion* is an atom that has lost or gained electrons, and thereby becomes charged (either + or –). An ion that has *gained* an electron will acquire an excess *negative* electric charge, and the ion formed by *losing* an electron will have a *positive* charge. These ions are very important in our life processes, and we would all die without them. (More about that later.)

Ions act differently than uncharged atoms. Think of them as being like magnets + and – , and other atoms as being like non-magnets. Does that start to explain how molecular formation can happen?

The "Abbreviated" Periodic Table of Elements provides some information about the formation of ions. As we discussed before, Group I elements can lose *one* electron, and Group II elements can lose *two* electrons. Guess what Group III elements can lose? *Three* electrons!

So, when atoms in Groups I, II, and III become ions, they will have a _____ charge.

Groups IV, V, and VI elements usually *share* electrons with other elements, and don't form ions. These elements will have a neutral charge.

Group VII elements attract and *gain one* electron. They can form ions, and they will have a _____ charge.

1. What is the electrical charge of an electron? _____

2. Fluorine is in Group VII. What would be the electrical charge of a fluorine (F) atom that gained an electron? _____

3. What would be the electrical charge of a sodium (Na) atom that lost an electron? _____

4. Refer to your "Abbreviated" Table of Elements, and determine whether the following atoms would form ions by losing or gaining electrons.

	Lose Electrons	Gain Electrons	Ion Electrical Charge
Cl			
Be			
Al			
Mg			
H			

CHEMICAL BONDS

Chemical bonds hold molecules together in living organisms. Furthermore, energy for the biological processes in living things is provided by the breaking and forming of chemical bonds in the molecules that drive our metabolism.

Two important types of chemical bonds are produced by the interactions between the electrons of different atoms: ionic bonds and covalent bonds.

IONIC BONDS

What do you think would happen if a + ion of lithium got very close to a − ion of fluorine? (Remember to think about ions as behaving like the + and − ends of magnets.)

The atoms of some molecules are held together by the attraction between oppositely charged ions. That force of attraction is called an *ionic bond*.

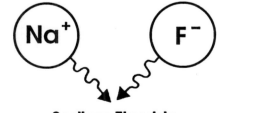

Sodium Fluoride
to prevent tooth decay

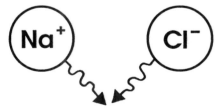

Table Salt

Ions are essential in many of the physiological processes of organisms such as nerve conduction and muscle contraction. Because ionic bonds are easily broken by water, the ions can be dissolved and moved throughout the fluids of an organism and delivered to every part of the body.

COVALENT BONDS

Elements in Groups IV, V, and VI *share* electrons with other atoms when bonding. The atoms of some molecules are held together by this mutual sharing of electrons. This cohesive force is called a ***covalent bond***.

Covalent bonds are important to living organisms because they are strong enough to hold together the molecular structure of the organism. However, these bonds aren't so strong that they can't be broken and "remodeled" into other molecules that the organism needs.

The covalent bonds in glucose are strong enough to make wood (cellulose).

The covalent bonds in glucose have enough energy to "drive" our metabolic processes (cell respiration).

The covalent bonds in glucose can be "remodeled" into other organic molecules that the organism needs.

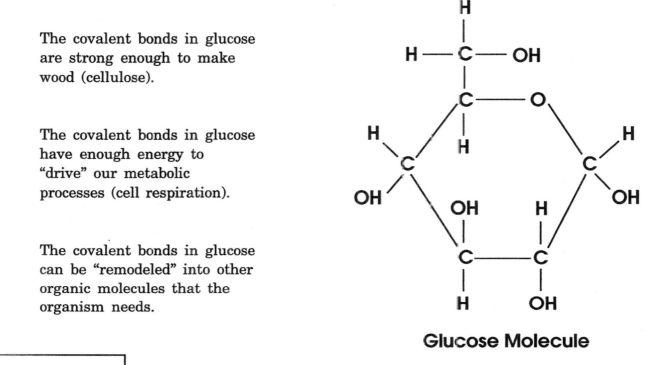

Glucose Molecule

? QUESTION

Based on what you know about elements in the Periodic Table, determine whether the following elements would interact to form ionic or covalent bonds.

Elements Reacting	Type of Bond
Na and Cl	
H and O	
H and Cl	
C and H	

ACTIVITY #4

"THE RULE OF EIGHT IN MOLECULAR FORMATION"

Remember: The elements in the Periodic Table of Elements are listed in vertical *groups* that have Roman numbers I through VII plus a group called the "inert gases." We have been discussing each group # with a special description of the behavior of electrons in the atoms of that group.

NOW

Whenever chemists observed atoms reacting with each other, they noticed that the *sum of their group numbers* usually equalled eight, or multiples of 8 (16, 24, 32, 40, etc.). This is called **"The Rule of Eight."** Let's see how this "Rule of Eight" works in understanding *chemical formulas.*

Chemical formulas are written to show how many atoms of each element are bonded together to make a molecule of a substance. For example $C_6H_{12}O_6$ is the chemical formula for one molecule of sugar.

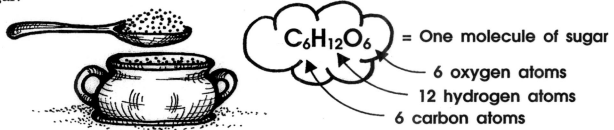

= One molecule of sugar

— 6 oxygen atoms
— 12 hydrogen atoms
— 6 carbon atoms

Let's look at water: H_2O. By looking at this formula, you know that it takes _____ hydrogen atom(s) to bond with _____ oxygen atom(s).

H_2O

Now, add the group numbers. Hydrogen has a group # of I (1). Oxygen has a group number of VI (6). Two hydrogens (1 + 1) and one oxygen (6) equals a total of 8. *"The Rule of Eight!"*

Try another simple one—table salt (sodium chloride): NaCl. The formula tells you that it takes _____ sodium atom(s) to bond with _____ chlorine atom(s). Add the group numbers: The group # of sodium is I (1) and the group # of chlorine is VII (7).

$$1 + 7 = 8. \text{ There's that } \textit{"Rule of Eight"} \text{ again!}$$

Try this one on your own. Go back to the sugar molecule: $C_6H_{12}O_6$.

6 atoms of carbon (group # _____) 6 x _____ = _____

12 atoms of hydrogen (group # _____) 12 x _____ = _____

6 atoms of oxygen (group # _____) 6 x _____ = _____

(Sum of all group numbers)

Total = _____

Is the total of _____ a multiple of 8? _____ *"The Rule of Eight"* again!

1. Determine the sum of the group numbers for the atoms in each of the molecules below, and see if your answer still agrees with the "Rule of Eight."

	Sum of Group #'s	Rule of 8?
NH_3 (ammonia)		
$CaCl_2$ (calcium chloride)		
Al_2O_3 (aluminum oxide)		

2. Let's see if you can use this "Rule of Eight" to write a chemical formula by determining how many *atoms* of one element would be necessary to combine with how many *atoms* of another element to form a particular molecule.

1 Mg + _____ Cl $\longrightarrow$ MgCl _____ (magnesium chloride)

2 H + 1 S + _____ O $\longrightarrow$ H$_2$SO_____ (sulfuric acid)

1 C + _____ H $\longrightarrow$ CH _____ (methane)

2 H + _____ O + 1 C $\longrightarrow$ H$_2$CO_____ (carbonic acid)

FINALLY

Chemists explain this "Rule of Eight" phenomenon by saying that all atoms (except for the inert gases) have *incomplete* outer electron shells. When atoms react with each other, each atom ends up with a compete outer shell of 8 electrons. This principle they call "The Rule of Eight" or "The Octet Rule."

There is much more to this idea, but we'll have to leave that to your further explorations into chemistry. For instance, if you consider the different molecules of gas in the air you breathe: H_2, O_2, or N_2, do their sum of group numbers equal 8? What's going on? Why?

For now, you will just have to be content with scratching the surface of chemistry. Understanding atoms, molecules, and the bonding of elements that make up all of the different substances that you are and you see on this planet will require a class in chemistry.

Summary Questions

1. Describe the hydrogen bond in water.

2. How is the hydrogen bond related to cohesion, adhesion, and capillarity?

3. What happens to the hydrogen bonds when water is boiled (evaporated)?

4. What is pH?

5. Why is pH important to living organisms?

6. What is a buffer?

7. How is the presence of a buffer important to living organisms?

8. Define osmosis.

9. How is osmosis important to plants?

CHEMISTRY CAPERS

Life first evolved in water, and most life on this planet lives in water. Furthermore, plant and animal forms are 70% to 90% water. These facts clearly establish the importance of this substance to all living beings. Water is rare in the universe, and because of this, life is also rare.

In this lab we will investigate some of the unique chemical properties of water: the most essential substance for living organisms.

ACTIVITIES

ACTIVITY #1

"PROPERTIES OF WATER"

In this Activity you will investigate some of the physical properties of water. A more thorough discussion of water's characteristics can be found in your textbook.

HYDROGEN BONDS

When two hydrogen atoms react with an oxygen atom to form water, there is an *unequal sharing* of electrons among the atoms. This creates a slightly negative charge on the side of the water molecule that has more of the "electron cloud."

Water molecules act like a bunch of magnets holding on to each other by the attractions between the + and − ends.

The force created by those attractions is called a **hydrogen bond**, and it is the secret to all of water's special properties.

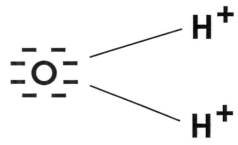

? QUESTION

1. As water falls from the clouds, what force keeps the water in drops?

2. In order for the liquid water to evaporate and become steam, heat must be added. In a pan of boiling water, what bond is being broken by the heat of the stove?

3. So, if heat is required to evaporate water, then what is released when water condenses?

4. On a calm, but rainy day, the temperature rises slightly when it starts to rain. Explain.

WATER ADHESION

Adhesion occurs when two or more different substances are stuck together as if by glue. Water has some interesting adhesion properties.

GO GET

1. A small container of water.

2. An eyedropper.

3. Four microscope slides.

1. Put several drops of water on one slide, then place the other slide directly on top of the wet slide.

2. Try to pull apart the glass slides *without* sliding them past each other.

3. Repeat this experiment with the two dry slides.

? QUESTION

1. How strong is the force of attraction between the two dry slides?

2. How strong is the force of attraction between the two wet slides?

3. What is the name of the force that holds the two slides together?

4. A freeze-dried anchovy is fairly easy to break between your fingers. Yet, when the fish is allowed to sit in water for a while, it only bends with the same effort. Explain.

5. Based on your answer to #4, what is one important role of water in living organisms?

CAPILLARITY

GO GET

1. A glass capillary tube.

2. A small container of water.

NOW

Hold the capillary tube vertically between your fingers, and put the bottom end just below the surface of the water.

The process you have just observed is called *capillarity*.

1. What happened when you did this experiment?

Capillarity

2. Draw a simple sketch of the results.

3. Explain your results. What is it about water that makes it do this? (Be specific about the forces of attraction.)

4. How is this property of water important to plants?

HEATING PROPERTIES OF WATER

The difficulty of heating water reveals the strength of the hydrogen bonds and the *temperature stabilizing role* of water within living organisms.

When a substance (like water) is heated, the rate of temperature increase depends on how easy it is for heat energy to increase the speed of molecular motion in that substance. If it doesn't require much heat to increase the motion, then we say that the substance is "easy to heat up."

You can apply this idea to the heating of water, and answer the question: ***"Are the hydrogen bonds between water molecules strong enough to make water a substance that is hard to heat up?"***

GO GET

1. A hot plate.

2. A chunk of metal.

3. A beaker (the 100-ml size is best).

4. A container of water.

5. Tongs to remove weight.

The cooling rate of an object is directly related to the amount of heat energy absorbed by that object. If an object cools quickly, then it didn't absorb much heat energy to start with.

1. Weigh the piece of metal.

2. Put an amount of water equal to the weight of the metal object into the beaker.

3. Put the piece of metal into the water, and heat the container until it just begins to steam.

4. Immediately remove the piece of metal from the beaker, and put both the beaker of water and the piece of metal onto the table. (Your results are more accurate if you pour the hot water into a room temperature beaker at the same time you put the metal on the table.)

5. Repeatedly touch both the water and the piece of metal until both are approximately the same temperature. Keep track of how long it takes for each to cool.

    Time for the water to cool = _____
    Time for the metal to cool = _____

? QUESTION

1. Which substance cooled the slowest? (circle your choice)

    Metal   or   Water

2. Which substance would require more heat energy to heat it up? Remember that the amount of heat given off by a substance equals the amount of heat absorbed by that substance when it was heated. (circle your choice)

    Metal   or   Water

3. We know that sitting in 70°F water is more chilling to the body than sitting in room air at 70°F. Explain why.

    If you were made of metal, would it be more chilling or not? Explain.

4. Based on your experimental results, what do you conclude about the importance of the hydrogen bond on the heating up of water?

5. Is water a temperature-stabilizing substance for living organisms?

EVAPORATION OF WATER

Heating water from 0° to 100°C just "shakes" the molecules. As you continue to heat water at 100°C you start "breaking" the hydrogen bonds. The **heat of vaporization** is the amount of energy needed to break hydrogen bonds. It can be measured if you know how much heat is used to boil away 10 ml of water already at 100°C.

GO GET

1. A hot plate.

2. Three equal-size beakers (the 250-ml size is best).

3. A thermometer. *Be careful, please!* This equipment is fragile.

NOW

1. Fill Beakers Ⓐ and Ⓑ with the proper amounts of water, and put them in the freezer or into special "ice tubs" for Beakers Ⓐ and Ⓑ to stay cold.

   Beaker Ⓐ is the heat meter. It will be heating during the experiment. Each 1°C increase in Beaker Ⓐ means that 100 calories of energy has left the hot plate and entered each of the three beakers.

   Beaker Ⓑ is a comparison to see which takes more energy— heating water or boiling water.

   Beaker Ⓒ is the experiment. How much energy is required to break (boil away) all of the hydrogen bonds in 10 ml of water?

2. Prepare a large container of boiling water. *Use the special measuring pipettes for safely removing 10 ml of boiling water.* You must prepare Beaker Ⓒ at the last minute just before you start the experiment. Read on.

3. Preheat your hot plate at a setting that you know will boil water moderately. (*Not the highest setting!*)

4. It is important that you start all three beakers at *exactly the same time*, without time for the beakers to change temperature before heating on the hot plate. Record the starting time as soon as all beakers are on the hot plate.

Beaker Ⓐ

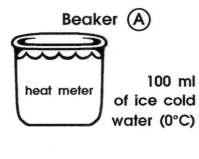

heat meter

100 ml of ice cold water (0°C)

Beaker Ⓑ

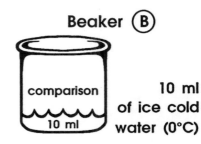

comparison
10 ml

10 ml of ice cold water (0°C)

Beaker Ⓒ

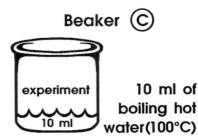

experiment
10 ml

10 ml of boiling hot water (100°C)

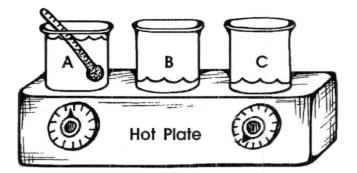

Hot Plate

5. *There are three events that you must record during this experiment.*

    a. How many minutes does it take for Beaker (B) to go from 0°C to 100°C (little bubbles form at the bottom)?

        _____

    b. How many minutes does it take for all of the water in Beaker (C) to boil away?

        _____

    c. Record the temperature in Beaker (A) at the exact time when all of the water finally boiled out of Beaker (C).

        _____ °C in Beaker (A)

? QUESTION

1. What takes more energy? (circle your choice)

    To heat 10 ml of water   or   To boil away (*evaporate*) 10 ml of water that
        from 0°C to 100°C.   is already at boiling temperature (100°C).

2. Based on your results, how effective is the evaporation of water (sweating) at removing excess heat from your body?

3. Beaker (A) (Heat Meter) tells us how much heat energy left the hot plate and entered the Experiment Beaker (C). One calorie of energy is the amount of heat required to increase the temperature of 1 ml of water 1°C. How much did the temperature change in Beaker (A)? _____ °C. (Subtract the starting temperature if it was above 0°C.)

4. Now, calculate the total calories of heat recorded by the Heat Meter (Beaker (A)) considering both the amount of water in it and the temperature change of the water.

        _____ calories

5. The answer to question #4 is the amount of energy necessary to vaporize 10 ml of water in Beaker (C). How many calories would be required to evaporate only 1 ml of water?

        _____ calories

6. What does this experiment tell you about calories, exercise, and sweating?

ACTIVITY #2

"WHAT THE HECK IS pH?"

HYDROGEN IONS

An ion is an atom that has lost or gained electrons, and thereby has become electrically charged (either + or −). Ions act differently than uncharged atoms. (It's like comparing magnets with non-magnets.)

These ions are very important in our life processes, and we would die without them. Table salt is an example of two essential ions—sodium and chloride.

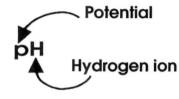

Of all the ions in your body, none is more important than the **hydrogen ion**, **H⁺**. The term *"pH"* refers to the concentration of H⁺ ions in water. Biologists are interested in H⁺ concentration because it affects chemical reactions so greatly.

A small change in this ion can dramatically affect life. Acid rain and acid stomach are two expressions of the concentration of H⁺ ions. Also, the blood of a human is so sensitive to H⁺ concentration that a small pH change from your normal of **7.4** can result in your death.

We monitor the pH of our fish aquariums and our swimming pools in order to avoid potential problems. In the case of the aquarium, we are trying to maintain a good environment for micro-organisms, whereas, in the swimming pool, we are trying to prevent micro-organisms from growing.

? QUESTION

1. Knowing how important pH can be to living organisms, what would be the effect of acid rain on the ecosystem?

2. What does the "p" stand for in the term pH?

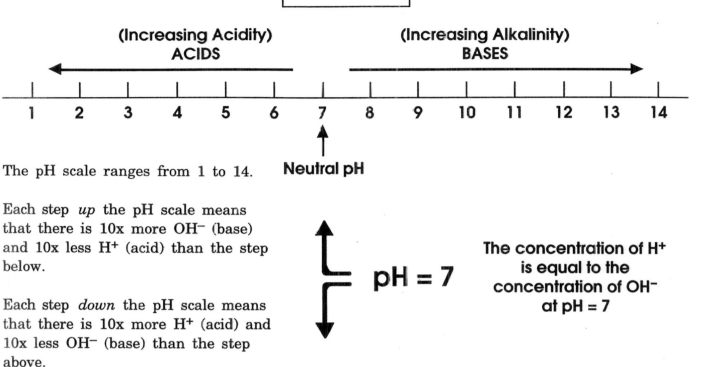

pH SCALE

(Increasing Acidity)
ACIDS

(Increasing Alkalinity)
BASES

1 2 3 4 5 6 7 8 9 10 11 12 13 14

Neutral pH

The pH scale ranges from 1 to 14.

Each step *up* the pH scale means that there is 10x more OH^- (base) and 10x less H^+ (acid) than the step below.

Each step *down* the pH scale means that there is 10x more H^+ (acid) and 10x less OH^- (base) than the step above.

$$pH = 7$$

The concentration of H^+ is equal to the concentration of OH^- at pH = 7

A water solution with a pH of 7 is **neutral** because the concentration of the H^+ ions (acid) is equal to the concentration of the OH^- ions (base).

If the pH is less than 7, then there are more H^+ (acid) ions than OH^- (base) ions, and the solution is called an **acid**.

If the pH is more than 7, then there are more OH^- (base) ions than H^+ (acid) ions, and the solution is called a **base**.

? QUESTION

1. pH is a measure of _____ ion concentration.

2. A pH of 3 would be . . . (circle your choice)

       Acid   or   Base

3. A pH of 11 would be . . . (circle your choice)

       Acid   or   Base

4. What is the relationship between H^+ and OH^- at a pH of 7?

5. How much more H^+ is in water at a pH of 3 when compared to a pH of 6?

6. How much more OH^- is in water at a pH of 11 when compared to a pH of 7?

77

HOW TO MEASURE pH

The pH can be measured with a machine or with special color indicators. A pH machine directly reads the H⁺ concentration and displays the pH on a screen. A color indicator is a special molecule that changes color at a particular pH level.

Follow the directions given by your instructor if you are using a pH machine. Otherwise, follow these instructions.

GO GET

A pH paper test kit.

NOW

1. Use the pH paper test kit to determine the pH of the *three unknown solutions* on the demonstration table.

2. Tear off a 1" strip of pH test paper, and squirt a drop on it from the test solution. Compare the color of the pH test paper to the pH color chart. That's the pH!

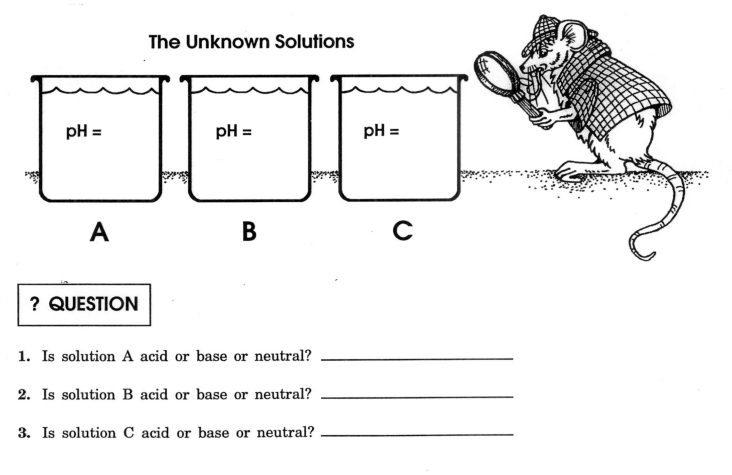

The Unknown Solutions

pH =   pH =   pH =

A   B   C

? QUESTION

1. Is solution A acid or base or neutral? _____

2. Is solution B acid or base or neutral? _____

3. Is solution C acid or base or neutral? _____

BUFFERS

A **buffer** is a chemical substance that can be added to water, and will make that solution *resist* a pH change.

GO GET	

1. 25 ml of *Sample X* in a small beaker.

2. 25 ml of *Sample Y* in a small beaker. } ***Mark both of the beakers!***

3. A dropper-bottle of phenol red.

4. A dropper-bottle of acid.

NOW

1. Put 5 drops of *phenol red* into each beaker.

   Phenol red is a pH color indicator. It turns *yellow* in *acid*, and it turns *red* in *base*.

2. Counting the drops, add acid one drop at a time until each beaker turns yellow. *Gently shake* the beakers after each drop in order to mix the acid into the test solution. (If either solution hasn't changed to yellow after 25 drops of acid have been added, then stop adding acid and assume that it will take many more drops to change the pH.)

? QUESTION

1. Which solution contains a buffer? (circle your choice)

   Sample X   or   Sample Y

2. How many *more* drops of acid did it take to change the buffered solution compared to the nonbuffered solution?

   _____ more drops.

3. Why do you think that one of the brands of aspirin is called "Bufferin"?

ACTIVITY #3

"MOLECULAR MOTION"

All atoms and all molecules move! They are bouncing off of each other at an incredible speed. (It's a good thing that O_2 and N_2 molecules are so small, because they would "sandblast" your skin if they were bigger.)

Chemists discovered that the speed of molecular motion is influenced by several factors, and we will investigate two of them. Also, we will look at a couple of special effects created by molecular motion.

CAN YOU SEE MOLECULAR MOTION?

Actually, molecules are too small to see. But a clever physicist calculated the energy in moving water molecules, and has determined that those molecules have enough energy to bump into and move some very small particles, like carmine dye. When viewed under a microscope, this movement can be seen.

GO GET

1. A slide and coverslip.
2. A compound microscope.
3. A drop of *carmine dye* particles.

NOW

1. Make a wet mount of the carmine particles.

2. Look at the *very very* smallest particles that you can see. Show your instructor.

3. The vibrating motion of these particles suspended in water is called **Brownian Motion**. (Named after guess who?) The tiny particles move whenever a water molecule (which you can't see) bumps into the carmine particle. Observe.

? QUESTION

1. When you watch *Brownian Motion* are you actually seeing *molecules* move? Explain.

2. What do you think would happen if you held a flame under the slide? Explain.

3. What would happen if you held an ice cube under the slide? Explain.

LIGHT MOLECULES VS. HEAVY MOLECULES

Methylene Blue dye has a molecular weight of 374. The purple dye *Potassium Permanganate* has a molecular weight of 158. If molecular weight makes any difference in the *speed* of motion, then we should be able to measure that difference.

You will need to work as a class for this experiment. Assign one member of your group to work with members from other groups.

GO GET

1. Two agar plates.

2. Put *one* crystal of *Methylene Blue* in the middle of the agar plate. (Your instructor may have you put a drop of Methylene Blue solution into one of the small depressions in the agar.)

3. Put *one* crystal of the *Potassium Permanganate* in the middle of the other agar plate. (If you are using solutions, put a drop of Potassium Permanganate into the other agar depression.)

NOW

1. Record the starting time for this experiment.

2. At 30 minutes and one hour, come back to the agar plates and measure the *diameter* of the spreading colors. Record your measurements in the chart below.

RATE of SPREAD (Diameter of Color)

Test Molecule	In 30 Minutes	In One Hour
Potassium Permanganate		
Methylene Blue		

3. Share the results with the other members of your lab group.

? QUESTION

1. Which molecules move faster? (circle your choice)

    Potassium Permanganate   or   Methylene Blue

2. What does this experiment tell you about the speed of movement of different-sized molecules?

ACTIVITY #4

"DIFFUSION OF WATER INTO AND OUT OF CELLS"

The movement of molecules from where they are in high concentration to an area where they are in low concentration is called **diffusion**. Because *all* molecules move, they also *diffuse*!

When a cube of sugar is put into a cup of coffee we know that the sugar will diffuse from where it is in high concentration (the cube) to where it is in low concentration (the hot coffee).

However, we don't normally think about what the water molecules are doing in that same cup of coffee.

The rules of diffusion apply to water concentration just like they do to the sugar. Water will diffuse from where it is in high concentration (the hot coffee) to where it is in low concentration (the sugar cube).

We can observe the effects of water diffusion in living cells. The diffusion of water through a cell membrane is called **osmosis**. This is a very important process involving the movement of water throughout the cells of all living organisms.

 GO GET

1. A slide and coverslip.
2. A leaf from the *Elodea* plant.
3. An eyedropper of salt water.

NOW

1. Make a wet mount of the *Elodea* leaf and look at the leaf cells under high power. Draw a picture of the distribution of chloroplasts within a typical cell.

2. Work with another lab group. One group is to leave their normal *Elodea* slide under the microscope. The other group is to add salt water to their *Elodea* slide.

3. Put one drop of salt water at the edge of the coverslip. Use a piece of tissue paper on the opposite side of the coverslip to absorb and pull the salt water across the slide. The salt water will soon surround all the leaf cells.

4. Wait 10 minutes. Look back and forth between the two microscopes. Draw another picture of the distribution of chloroplasts within a typical cell in salt water. *Be sure to clean the microscope stage if you got any salt water on it.*

Elodea Cell

Salt Water Added

Let this picture represent one *Elodea* cell surrounded by salt water. The small circles are the water molecules and the larger circles are the salt molecules.

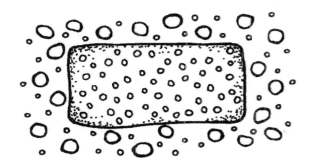

Salt Molecule ⬭

Water Molecule ○

1. Where is the *water* in high concentration? (circle your choice)

    Inside of the cell   or   Outside of the cell

2. Where is the *water* in low concentration? (circle you choice)

    Inside of the cell   or   Outside of the cell

3. In what direction will *osmosis* (water diffusion) occur? (circle your choice)

    Into the cell   or   Out of the cell

4. Explain what you saw in your second picture of the *Elodea* leaf cells. (What is the membrane surrounding the grouped chloroplasts? What happened to the central vacuole?)

5. When you eat a lot of highly salted food (a bag of potato chips, for example), what happens?

Why?

6. People with high blood pressure or heart problems are told to be careful about their intake of salt. Why?

FINALLY

Under normal conditions root hair cells are relatively low in water molecules and high in other kinds of molecules (cell salts and nutrients). *Draw arrows* to show water movement between the soil water, the root hair cell, and the water transporting tube.

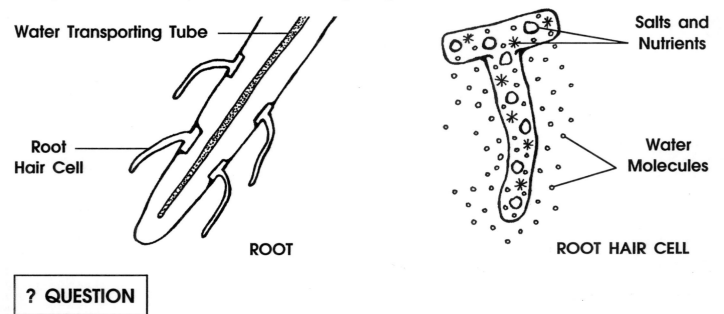

? QUESTION

1. What would happen to the water movement if you put salt or a lot of fertilizer in the soil?

2. Explain how the properties of water investigated in Activity #1 and Activity #4 assist a plant in the movement of water from the soil, into the root hair, and throughout the structure of the plant.

OPTIONAL

Water is a most unusual substance. It has some strange changes in volume at different temperatures. Try this surprising experiment.

1. Very accurately mark the water level in a beaker of cool water.

2. Heat the beaker of water to just before steaming (when bubbles start to appear) and record the water level.

3. Refill the beaker with ice cubes and enough water to match your first water level mark. (Push all the ice below the water surface.)

4. Allow the ice to melt, and record the water level.

5. Compare your results. Surprised?

ENZYMES

Summary Questions

1. What kind of molecule is an enzyme?

2. What is the key feature responsible for enzyme activity?

3. Discuss how enzymes are related to the "energy of activation" in a chemical reaction.

4. What is an experimental "control," and why must it be included in all experiments?

5. What effect does overheating have on the potato enzyme?

6. What could happen if an animal's enzymes became overheated in a desert climate?

7. What effect does an acid solution have on the potato enzyme?

8. What effect could excess acidity in the soil or rain have on the enzymes of living organisms?

ENEYMES

INTRODUCTION

Living cells need to build molecules and break molecules. **Enzymes** are special proteins made by the cell, and they greatly speed up the chemical processes of molecular making and breaking.

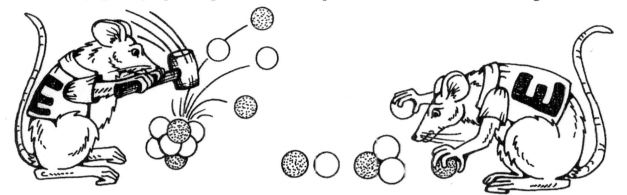

If cells did not have enzymes, then high amounts of heat energy would be required to perform the necessary chemical reactions. That quantity of heat would destroy most structures in the cell, and therefore would be detrimental to the life processes of the whole organism.

Because enzymes are so important to life, our genetic systems have evolved special genes to control their production. In fact, biochemists have discovered that the only differences between one species and another species are the *kinds of enzymes* produced and the genetic sequence of that production.

This lab is designed to help you develop a general understanding of how enzymes operate, and appreciate some of the environmental factors that influence their operation.

ACTIVITIES

ACTIVITY #1

"ENZYME ACTION"

THE ROLE OF HEAT IN CHEMICAL REACTIONS

Heat energy can be used to break the chemical bonds holding big molecules together, producing smaller molecules. Heat also speeds up molecular motion, thereby increasing the frequency of molecular collisions. Collisions of small molecules can result in the spontaneous formation of larger molecules.

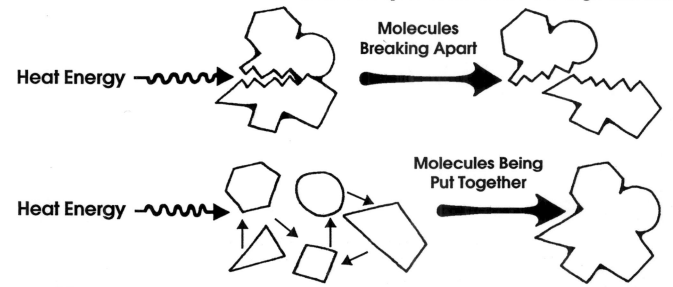

Because of these two properties of heat, molecules can be created and restructured into new substances. However, a problem arises when you need to do both jobs at the same time. In a big chemical factory, reactions can be separated so that the heat needed to drive one process won't interfere with the other processes. Because of its structure and size, a cell isn't able to do this. Therefore, a cell needs another solution for the control of biochemical events—*enzymes*.

? QUESTION

1. When you add *heat* energy to a chemical system, what happens to the speed of molecular reactions?

2. But, if too much *heat* energy is added to a system, what happens?

ENZYME SHAPE

Enzymes have unique **shapes** and **reactive spots** that allow them to break apart molecules or put together molecules without using very much heat. The special shape of a particular enzyme will attract and hold two molecules close to each other until they chemically bond. Or, a slightly *different* enzyme will break a large molecule into smaller molecules.

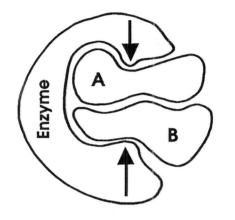

$$A + B \longrightarrow AB$$

OR

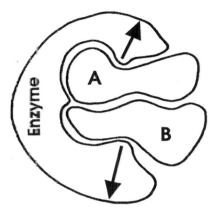

$$AB \longrightarrow A + B$$

? QUESTION

1. What do you think creates the unique shape of an enzyme?

2. What would happen to a normal chemical process in your cells if the shape of the enzyme controlling that process was changed by an environmental toxin?

ENZYMES AND ENERGY OF ACTIVATION

All reactions need a "push" to get started. This push is called the *energy of activation*. Sometime the "push" involves quite a lot of energy. If the cell was dependent only on heat to supply the energy of activation, then the cell would soon *overheat*, and molecules would be *destroyed* by that heat energy. Enzymes radically decrease the "push" required to start a reaction.

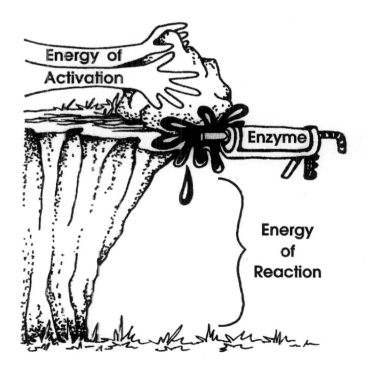

Another way of thinking about enzyme function is to imagine that they are like the grease in the picture. And, like grease, enzymes can be used over and over, because they aren't actually changed by the chemical reaction.

Any substance that speeds reactions and is not used up during the process is called a *catalyst*. Enzymes are catalysts.

1. What do enzymes change about the *amount* of energy required to make chemical reactions occur?

2. Are enzymes destroyed or used up during the chemical reaction?

3. What is another term for a substance that speeds a reaction but is not part of the end product?

GENES AND ENZYMES

There are many thousands of different chemical reactions that occur during the life of an organism. Each of these reactions is controlled by its own enzyme.

Obviously, every organism must possess the ability to manufacture these enzymes. Each enzyme is created under the direction of a cluster of **genes**, and you have many, many genes (30,000 or so in a human). You need lots of genes because many different enzymes are necessary to perform all of the biochemical processes that make up your life as a complex organism.

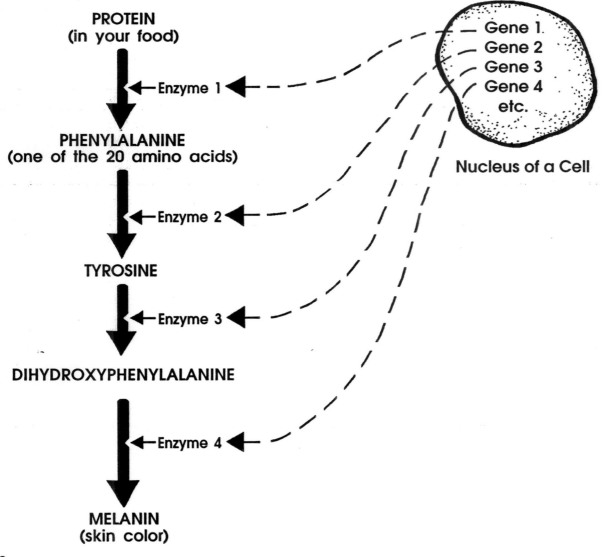

PROTEIN
(in your food)

← Enzyme 1 ←

PHENYLALANINE
(one of the 20 amino acids)

← Enzyme 2 ←

TYROSINE

← Enzyme 3 ←

DIHYDROXYPHENYLALANINE

← Enzyme 4 ←

MELANIN
(skin color)

Gene 1
Gene 2
Gene 3
Gene 4
etc.

Nucleus of a Cell

90

1. Why does your body need so many different enzymes?

2. What component in the nucleus is responsible for synthesizing a particular enzyme?

3. What kind of molecule is an enzyme?

4. If a gene was changed by "mutation," then what do you think would happen to the shape of the enzyme built by that gene?

5. Based on your answer to question #4, what would be the consequence to the organism affected by those "mutations"?

6. 98.3% of all human genes also occur in the chimpanzee. What does this suggest about the similarity of chemistry in these two species?

7. There is only a 0.4% genetic difference among all humans. What does this suggest about the similarity of chemistry among humans as compared to that between humans and chimpanzees?

ACTIVITY #2

"EXPERIMENTAL DESIGN"

In science, we try to make sure that the work done during the day gets us a little closer to understanding what is really going on in the world. *Experimental Design* is the planning of an experiment so that we get a *yes* or a *no* answer to a question we are asking.

THE EXPERIMENTAL CONTROL

The scientific method begins with a possible explanation of something we are observing. That explanation is a called the *hypothesis*. It generates the questions we are trying to answer in an experiment.

For example, we may think that adding more water to a lawn will make it grow faster, and we could design an experiment to test that hypothesis. In the "experimental design" we might decide to water the lawn twice as much as normal rainfall.

Let's suppose that we do this experiment for a month, and we observe that the lawn actually does grow quite a bit faster. Can we safely conclude that we have tested the hypothesis?

What if someone were to point out that the weather was warmer than normal during our experiment and that is why our *"experimental"* lawn grew faster than normal?

Or suppose that someone else said that our lawn really didn't grow any more than her lawn which was watered only by rainfall during the same month?

Because we can't answer these questions, our "experimental design" has failed. Why?

We did not control our experiment! An experimental *control* is a duplicate procedure that is set up exactly like the experiment *except* that the *factor being tested* (more water) *is left out*. So, in our experiment, we should have monitored a nearby lawn that received only normal rainfall during the same warm month. Then we could compare the growth rate of that "control" lawn to our "experimental" lawn which was receiving extra watering. The results from observing our control lawn would have answered both questions about our experiment's design.

All experiments need a control!

A SIMPLE ENZYME REACTION THAT WE WILL ALL DO TOGETHER

Work in groups of 3 or 4 people.

In simple words, an enzyme speeds up the conversion of a substance—called a *substrate*—into a *product*.

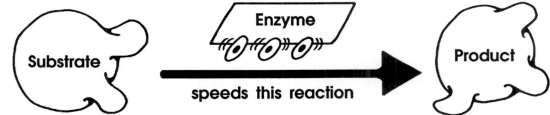

92

Potato juice has an enzyme that will change a colorless substrate (called catechol) into a yellow-brown product (called quinone). Forget the fancy names; we will call them substrate and product.

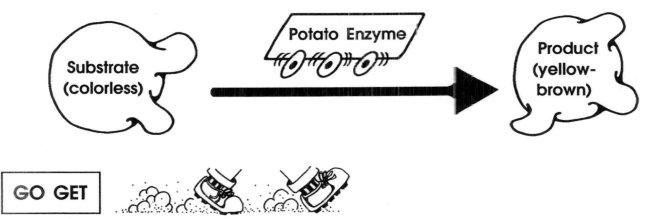

GO GET

1. Two test tubes. *Rinse them out* with tap water. They may be contaminated from the last class.

2. Colorless substrate.

3. Potato extract enzyme.

NOW

1. Fill each tube half full with distilled water.

2. Add 10 drops of the colorless *substrate* to one of the test tubes and shake the tube. This is the "control" for the experiment. It does not have the enzyme.

3. Add 10 drops of the colorless *substrate* plus 10 drops of the potato extract *enzyme* to the other test tube and shake the tube. This is the "experimental" tube. It has the enzyme.

? QUESTION

1. Why are we adding the enzyme to the second test tube?

2. Did you observe any *product* forming? (**Remember:** A yellow-brown color means that the product has been made.)

   If so, how long did it take to form the product?

3. What observations can you make about the first test tube?

4. Why is the first test tube necessary for the validity of your experiment?

5. What are your conclusions?

ACTIVITY #3

"FACTORS AFFECTING ENZYME ACTION"

This part of the lab will help you learn how to design experiments and to observe some of the environmental factors that can affect enzyme action.

NOW IT'S YOUR TURN

1. Divide into groups of 3 or 4 people.

2. Your group is to design an experiment with a control to answer *two* of the six questions below.

   (Your instructor may require that you test more than two of the questions. Be prepared to do so.)

3. Once your group has worked out an experimental design for the two questions, check with your lab instructor for possible suggestions. (For example, if you didn't time the reaction in Activity #2, you must do it in these experiments.)

4. Your group is to perform the experiments you design.

5. Your group will report its results and conclusion to the rest of the class later in the lab period. Fill out your lab report. Be sure your lab report is complete.

THE SIX ENZYME QUESTIONS

#1   Is the speed of the reaction influenced by the *temperature* of the environment?

#2   Is the speed of the reaction influenced by the *amount of enzyme* in the environment?

#3   Is the speed of the reaction influenced by the *amount of substrate* in the environment?

#4   Is the speed of the reaction influenced by the *pH* of the environment?

   In Activity #2, you ran the experiment at pH = 7. Use the prepared acid and base solutions instead of water to test the effect of pH on the enzyme. Leave the enzyme for 10 minutes in the test tube half-filled with acid or base. Then add the substrate to see if the enzyme is still active.

   Be very careful when handling acids or bases. Wash your hands or eyes immediately if any solution touches them.

#5   Are there natural substances, such as phenylthiourea, that can *inhibit* enzyme action?

   Use the same approach as in #4 leaving the enzyme in the tube of phenylthiourea for 10 minutes before adding the substrate. *Be careful.* Phenylthiourea is a poison.

#6   Are enzymes destroyed by *high heat*?

LAB REPORT

Question # _____

Hypothesis:

Experimental Design:

Results:

Conclusions:

Question # _____

Hypothesis:

Experimental Design:

Results:

Conclusions:

1. Sometimes raw vegetables and fruits are put into the refrigerator to slow the "browning" effect. How can you explain this based on results of experiments during this lab?

2. Sometimes lemon juice is put on raw vegetables or fruits to slow the "browning" effect. How can you explain this based on results of experiments during this lab?

3. When fruits or vegetables are left exposed to the air after they have been boiled, they don't turn brown. How can you explain this based on results of experiments during this lab?

Summary Questions

1. Write the equation for respiration.

2. What is an endotherm? Give two examples.

3. What is an ectotherm? Give three examples.

4. Describe the differences in metabolic rate of ectotherms and endotherms at different environmental temperatures.

5. Do plants do respiration?

   How do you know?

RESPIRATION

Cellular *respiration* is the process in living organisms that extracts electron energy from the chemical bonds in *food* (organic molecules), and converts that energy into a more useful form of energy (called **ATP**) to run cell activities. This cell process uses oxygen and produces carbon dioxide. The complete equation is:

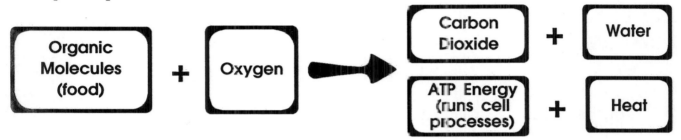

Respiration occurs inside the *mitochondria*, which are cellular organelles in both plant and animal cells. Refer to your textbook for the structural and functional description of this organelle.

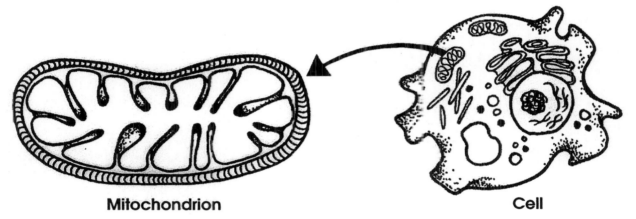

Mitochondrion   Cell

During this lab we will investigate some aspects of cellular respiration including the effects of environmental temperature on the rate of respiration in *endotherms* (internally heated animals) and *ectotherms* (externally heated animals).

ACTIVITIES

ACTIVITY #1

"HEAT PRODUCTION DURING RESPIRATION"

The Second Law of Thermodynamics states that heat is released whenever any form of energy is transformed into another form.

Since *respiration* is described as the conversion of food energy into usable energy for the cell, we should be able to observe heat being given off during the process.

OBSERVE

Two experimental containers were set up yesterday. One of the containers was filled with *dead seeds* killed by boiling, and the other container was filled with *live seeds*. Record the temperature of each container.

Temperature of live seeds = _____

Temperature of dead seeds = _____

These seeds demonstrate the basic respiration process that is going on in all living organisms.

Live Seeds

Dead Seeds

? QUESTION

1. What does the equation for respiration say about *heat*?

2. What does this experiment suggest is occurring in live seeds and not in dead seeds?

3. What would happen to the respiration process in the container of live seeds if we pumped the oxygen out?

What would happen to the temperature in that container?

ACTIVITY #2

"RESPIRATION IN AN ENDOTHERM"

Mice are **endotherms**. That is, they get most of their heat from *inside* their own body (*endo* means inside). Cellular respiration generates the heat that keeps these animals warm. (Refer to the Equation for Respiration on the first page.)

During this Activity you will monitor the ***rate of respiration*** (also called ***metabolic rate***) in a mouse. In addition, you will investigate the *influence of environmental temperature* on the mouse's rate of respiration by comparing a mouse in a *cold environment* with a mouse in a *warm environment*.

Later, in Activity #3, you will compare the differences between an endotherm (mouse) and an ectotherm (frog).

HOW TO HANDLE MICE

Mice should be *picked up by their tail* and immediately *rested on your hand,* and then marched into the Metabolic Cage.

Do not grab them.

Grabbing scares the hell out of them, and they may bite you or pee on you because of that fear.

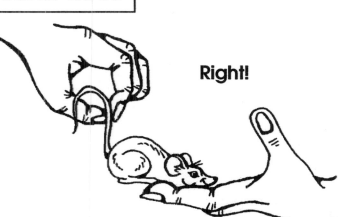

Right!

Wrong!

Also, *don't play with the mice* (on table tops, etc.) because there is a possibility of them getting loose on the floor.

These are professional mice. They work several years for us, and we treat them very well. So, please be careful.

EXPERIMENTAL APPARATUS

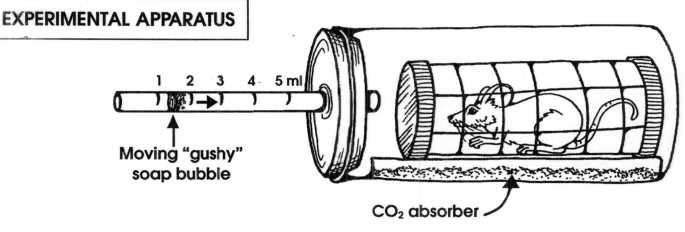

Moving "gushy" soap bubble

CO_2 absorber

EXPERIMENTAL DESIGN

The basic question is: ***What effect does environmental temperature have on the metabolic rate of an endotherm (mouse)?***

Do this experiment at two temperatures: Room Temperature and Packed in Ice.

ROOM TEMPERATURE

STEP 1 Weigh the wire cage part of the chamber: _____ grams.

STEP 2 Go get your mouse, and put it into the wire cage. Then weigh the cage with the mouse in it.

Cage + Mouse		**Cage**		**Weight of Mouse**
_____ g	−	_____ g	=	_____ g

STEP 3 Put one tablespoon of CO_2 absorber (soda lime) into the trough at the bottom of the Metabolic Rate Chamber.

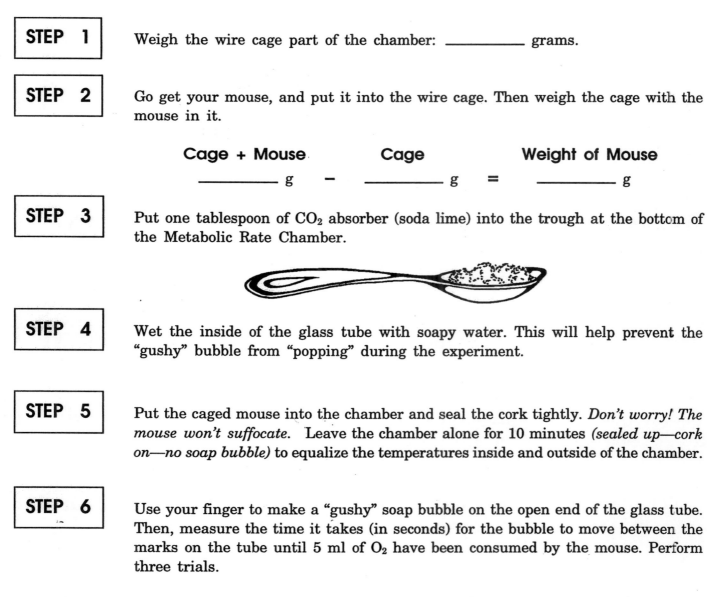

STEP 4 Wet the inside of the glass tube with soapy water. This will help prevent the "gushy" bubble from "popping" during the experiment.

STEP 5 Put the caged mouse into the chamber and seal the cork tightly. *Don't worry! The mouse won't suffocate.* Leave the chamber alone for 10 minutes *(sealed up—cork on—no soap bubble)* to equalize the temperatures inside and outside of the chamber.

STEP 6 Use your finger to make a "gushy" soap bubble on the open end of the glass tube. Then, measure the time it takes (in seconds) for the bubble to move between the marks on the tube until 5 ml of O_2 have been consumed by the mouse. Perform three trials.

_____ seconds	_____ seconds	_____ seconds
Trial 1	**Trial 2**	**Trial 3**

? QUESTION

1. Food + O_2 $\longrightarrow$ CO_2 + H_2O. During respiration a mouse will consume O_2, and CO_2 will be produced in its place. If no CO_2 absorber had been used in your experiment, would you have seen a change in air *volume*?

2. If you use a CO_2 absorbing substance in the Metabolic Rate Chamber, then what happens to the CO_2 that is produced during respiration?

3. Now, with the absorbing substance in the chamber, what happens to the *air volume* during your experiment as the O_2 is consumed during respiration?

PACKED IN ICE

If ice is packed around a Metabolic Rate Chamber like the type we are using, the temperature inside will stabilize at 5°C.

This cold air temperature *will not harm* the mouse as long as the mouse is removed before 45 minutes. Our experiment will take less than 20 minutes.

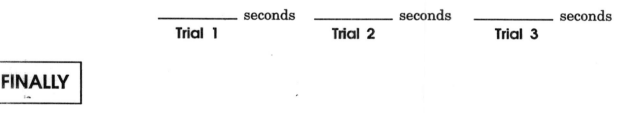

STEP 1

Now perform the "Packed in Ice" experiment. Let the chamber equalize the temperatures inside and out for 10 minutes before applying the "gushy" soap bubble.

STEP 2

After 10 minutes, apply a "gushy" soap bubble and perform the three separate measurements of the rate of respiration.

_____ seconds   _____ seconds   _____ seconds
Trial 1   **Trial 2**   **Trial 3**

FINALLY

Disassemble the chamber, carefully returning your mouse to its home, and dump all CO_2 absorber and feces into the special waste jar. Don't wash the apparatus unless you are told to do so. The chamber must be dry for the next lab class.

Wash your hands!

RESPIRATION CALCULATIONS

You must convert the mouse's O_2 consumption to an hourly metabolic rate. This is accomplished by dividing the bubble time (in seconds) into 3,600 (the number of seconds in one hour). That number is to be multiplied by 5 (5 ml of O_2 used in each trial).

CALCULATION 1

Calculate the *average* time of the three trials at room temperature.

5 ml O_2 consumed in _____ seconds (average time)

CALCULATION 2

Based on Calculation 1, how much O_2 would your mouse consume in one hour?

$$\frac{3,600}{\text{Calculation 1}} \quad x \quad 5 \quad = \quad \text{_____} \quad \text{ml } O_2 \text{ consumed in one hour}$$

CALCULATION 3

In order to have a metabolic rate that can be compared with an animal of different weight, we must correct the calculations considering the mouse's weight.

$$\frac{\text{Calculation 2}}{\text{Weight of Mouse}} \quad = \quad \text{_____} \quad \text{ml } O_2 \text{ per hour per gram of weight}$$

NOW

You have finished the calculations for room temperature. Record your answer below. Repeat the three calculations for "Packed in Ice," and record your answer below.

Metabolic rate of your mouse   =   _____   ml O_2 per hour
at room temperature (20°C)   per gram of weight

Metabolic rate of your mouse   =   _____   ml O_2 per hour
packed in ice (5°C)   per gram of weight

1. Put a dot on the graph for each of the metabolic rate values in your experiment.

2. Draw a line between those two dots, and write the word *endotherm* on the line.

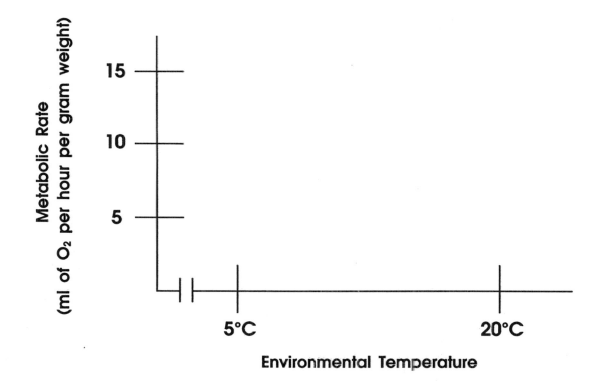

3. Check with other lab groups to see how your calculations compare with theirs.

ACTIVITY #3

"COMPARISON OF ENDOTHERM AND ECTOTHERM"

An **ectotherm** gets its heat from the environment (*ecto* means outside). The body temperature of an ectotherm is warm when the environment is warm, and the body is cooler when the environment is cold.

INFORMATION

The following results are taken from some experiments that measured the metabolic rate in a frog (*ectotherm*) of about the same size as your mouse.

	Metabolic Rate Packed in Ice (5° C)	Metabolic Rate at Room Temperature (20° C)
Frog #1	0.05	0.30 ml O_2 per hour per gram of weight
Frog #2	0.03	0.28
Frog #3	0.04	0.25

NOW

1. Calculate the *average* metabolic rate for the three frogs at each of the two temperatures.

2. Put a dot on the graph for each of the average values.

3. Draw a line between those two dots, and write the word *ectotherm* on the line.

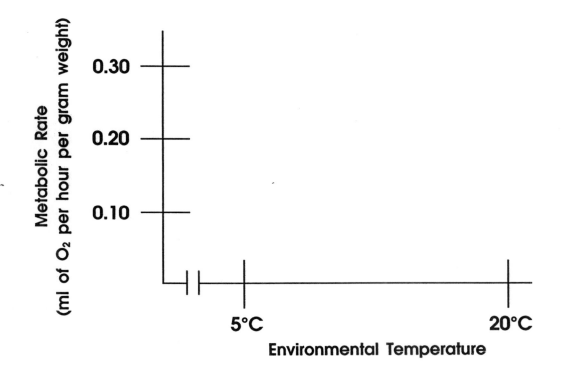

1. Which organism has the slowest rate of respiration? (circle your choice)

    Endotherm   or   Ectotherm

2. Which organism needs less food to survive? (circle your choice)

    Endotherm   or   Ectotherm

Explain why.

3. How much food does the *ectotherm* need compared to the *endotherm*? _____ %

4. Which organism would do better if the amount of food is very limited, but the environment is fairly warm? (circle your choice)

    Endotherm   or   Ectotherm

5. In what areas of the world would you expect to find ectotherms?

6. Which organism would do better in cooler environments where the food is plentiful? (circle your choice)

    Endotherm   or   Ectotherm

7. Will the organism in question #6 do fine in warmer environments if the food is plentiful?

Why or why not?

ACTIVITY #4

"FOOD DEMAND FOR HUMANS"

How much food does a human need to survive one hour of biology lab class?

We can borrow data from experimental research to help us estimate the amount of food that is required to support a human. Our calculations will be based on grams of sugar as the nutrient. Also, notice that the word *Calorie* is capitalized. When capitalized, this term represents 1000 times the value of a single calorie.

The Caloric demand for food varies greatly for a human depending on activity and environmental conditions. The energy demand might be as slow as 50 Cal per hour during sleep to as fast as 2,000 Cal per hour during extreme exercise. (Although that high rate of metabolism could be maintained for only about 2 minutes without total exhaustion.)

An average student in biology lab class uses about 100 Calories per hour as long as they aren't walking around all of the time.

INFORMATION

1. Assume a food demand of 100 Cal/hour for students.

2. A human gets about 3.85 Cal of energy from 1 gram of sugar.

? QUESTION

How many grams of sugar are required to "fuel" an average student during one hour of biology lab class?

_____ grams of sugar used in one hour

NOW

Weigh out that much sugar and show it to your lab instructor.

ACTIVITY #5

"RESPIRATION IN PLANTS"

The Respiration Equation states that CO_2 is produced as O_2 is used. If that is true, then we should be able to use CO_2 production as an indicator that respiration is occurring.

PHENOL RED TEST FOR CO_2

There is a very simple way to show changes in CO_2 level. Phenol red is a substance that turns yellow when CO_2 is added, and then turns back to red when CO_2 is removed.

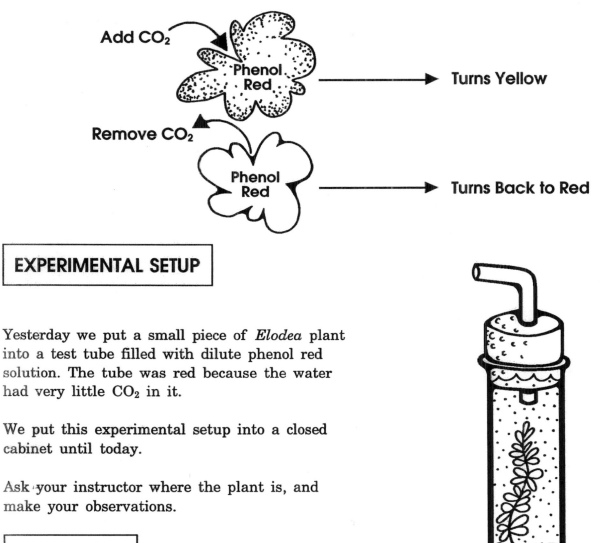

Add CO_2 → Phenol Red → Turns Yellow

Remove CO_2 → Phenol Red → Turns Back to Red

EXPERIMENTAL SETUP

Yesterday we put a small piece of *Elodea* plant into a test tube filled with dilute phenol red solution. The tube was red because the water had very little CO_2 in it.

We put this experimental setup into a closed cabinet until today.

Ask your instructor where the plant is, and make your observations.

? QUESTION

What do you conclude about plants in the dark?

Summary Questions

1. Describe the energy balance principle of your body.

2. Compare how many Calories are provided by 1 gram of carbohydrate, protein, and fat.

3. What is the underlying "secret" to the temporary success of the "Eat All You Want of This or That" Diet?

4. Why does the "Cut Something Out" Diet work so well?

5. What lifestyle change is guaranteed to create permanent weight loss and maintain balance with your food consumption? (**Hint:** see question #1.)

6. Discuss how brown fat operates with respect to fat metabolism.

7. What special problem in appetite control is created by the person who has a low amount of physical activity?

DIET ANALYSIS

Americans spend more money on high-Calorie fast foods and low-Calorie diet foods than anyone else in the world. In addition, more books have been written on topics related to body weight than any other health issue. The next time you are in the supermarket, notice how much shelf space is devoted to ailments of the digestive system. Either people have a headache from worrying about what they do eat, or they have a bellyache from what they did eat.

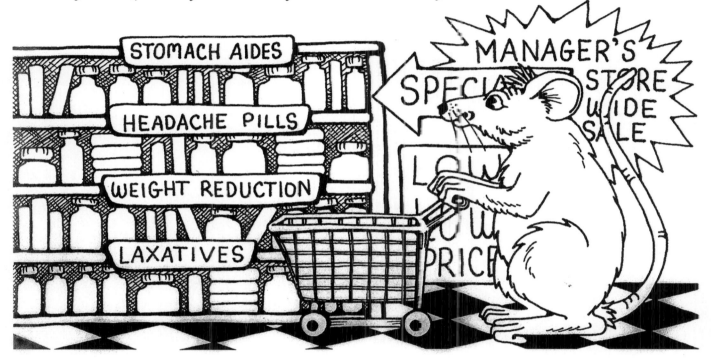

This week's lab will reveal some insights gained from using a scientific approach to the study of your diet.

ACTIVITIES

ACTIVITY #1

"ENERGY BALANCE OF THE BODY"

The energy balance of the body is a simple principle that explains weight changes. ***Remember: If energy input*** (food eaten) ***equals energy output*** (food metabolized for activity), ***then the body weight remains constant.***

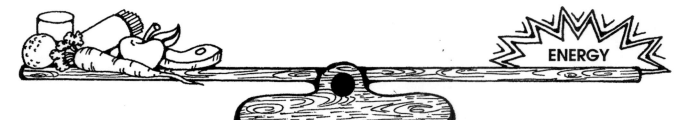

If more energy (food) goes into the body than is used by the metabolic processes, then your body weight will *increase* (fat storage). Likewise, if less energy goes into the body than is used, body weight will *decrease*.

ENERGY INPUT

The total chemical energy in a food can be determined by measuring the amount of heat given off when that food is burned. Several complications arise when human foods are analyzed. Some of the chemical energy in food cannot be digested. A primary component of plants is *cellulose* (cell walls). Our stomachs do not produce the enzyme to break this starch into sugars for energy. Termites and cows, however, do have this ability, which is why they can get more energy from eating plants.

Another complication in calculating the energy input for humans is that some of the digestible nutrients in our food aren't completely absorbed by the small intestine. Absorption varies from person to person. It may depend on the concentration of digestive enzymes or on the speed that foods are moved through the digestive tract. This is similar to other physiological differences in metabolism. For example, some people have inherited an ability to easily store fat. This may have been necessary for our ancestors. They needed to store fat during times of abundant food, and lived off that fat during periods of food shortage.

Although variations in physiology complicate the calculations of energy input from foods, the energy input values listed in the following table are accurate enough to begin the estimation of energy balance in your body.

Important Energy Estimates *
1 g of carbohydrate = **4** Calories of energy input
1 g of protein = **4** Cal of energy input
1 g of fat = **9** Cal of energy input
4200 excess input Calories = 1 pound gain in fat (9.3 x 454 g)
4200 deficit of Calories = 1 pound loss in fat

* These estimates are "rounded off" for easier calculation.

ENERGY OUTPUT

Energy output is the term used to represent all of the energy required to maintain the metabolic processes and activities of an organism. The most accurate and direct way of determining the amount of energy used by the body during an activity is to measure the **amount of heat given off**. Physics tells us that *heat is released whenever energy is transformed from one form into another* (such as nutrient energy into physical work). A technical problem with using this method of analysis is that the person must be inside an insulated container surrounded by a known quantity of water. The temperature of the water increases as heat is released by the person. Although this procedure is very accurate, it is also expensive and difficult.

Another method of measuring energy output is the **oxygen consumption technique**. If the oxygen requirement of a resting person is known, then that value can be compared to the increased amount of oxygen used during a particular physical activity. This approach is easier and less expensive than the heat method. Most general studies of energy expenditure are based on oxygen consumption.

1. One Calorie of fat provides more input energy than one Calorie of protein. (T or F)

2. One gram of fat provides more input energy than one gram of carbohydrate or protein. (T or F)

3. Assume that a person requires 1600 Cal to maintain constant weight. If this person eats no food (water only), then how much fat can they lose in a week? (Assume that only fat is being metabolized during the fast. Actually, your essential proteins and sugars are also used as an energy source during a fast, and their depletion can be a serious health threat.)

4. What important law of physics tells us that heat production is an accurate estimate of energy expended by the body?

5. You will be using several foods as examples during this lab. Let's see how you evaluate them before doing the Activities. Put a check mark if you think the food is high Calories, high protein, or high fat.

Food	High Calorie	High Protein	High Fat
Tuna Sandwich			
Whole Milk (8 oz)			
After School Snack:			
1 cup of peanuts			
6 crackers			
1 oz of cheese			
Cheeseburger			

6. Why does the burning of some foods provide an inaccurate Calorie value of that food for humans?

7. If the normal resting energy output is 80 Calories per hour and the energy output during moderate exercise is 200 Calories per hour, then what is the energy output for the exercise?

ACTIVITY #2

"NUTRIENTS IN FOOD"

Food Composition Tables are included in the appendix of most nutrition books. These reference tables give approximate amounts of carbohydrates, protein, and fat in the foods listed. On the last page of this lab there is an abbreviated table you can use for all the Activities.

BASIC CALCULATIONS

A tuna sandwich is the example we will use for calculating the amounts and percentages of nutrients in food. A sandwich is a mixture of ingredients, and the following information was obtained from the reference tables.

Tuna Sandwich	Carbohydrate	Protein	Fat
2 Slices of Bread	24 g	4 g	1.4 g
1 Tbsp. of Mayonnaise	trace	trace	11 g
Lettuce	trace	trace	trace
2 oz of Tuna	0 g	15 g	1 g
Totals	24 g	19 g	13.4 g

STEP 1 The calculations for nutrient analysis *based on weight* take some time but are simple to do. The first step is to determine the ingredients in a food. This was done for you in the tuna sandwich example above.

STEP 2 Use the nutrient information in the Food Composition Table at the end of this lab to determine the total weight (in grams) *for each category of nutrient* (carbohydrate, protein, and fat). This was done for the tuna sandwich.

STEP 3 Add the weights of all three nutrients (C + P + F). For the tuna sandwich the total weight of carbohydrate plus protein plus fat is 24 + 19 + 13.4 = 56.4 g.

STEP 4 Calculate the *percent* of each nutrient in the food using the following formula:

$$\frac{24}{56.4} = 42\% \ \textbf{C}$$

$$\frac{\text{(Weight of a Nutrient Category)}}{\text{(Total Weight of All Nutrients)}} \times 100 = \underline{\ \ ?\ \ } \%$$

$$\frac{19}{56.4} = 34\% \ \textbf{P}$$

$$\frac{13.4}{56.4} = 24\% \ \textbf{F}$$

117

Let's see if you can calculate the weight percentages of each nutrient in the entrées below.

Entrées	C	P	F
1. 8 oz glass of Whole Milk Totals			
Percentages by Weight			
2. After School Snack: 1 cup of peanuts			
6 crackers			
1 oz of cheese			
Totals			
Percentages by Weight			
3. Cheeseburger Totals			
Percentages by Weight			

ACTIVITY #3

"CALORIES FROM NUTRIENTS"

In Activity #2 you calculated the **weight percentages** of nutrients in a tuna sandwich. The **percentages based on Calories** of each nutrient is a more accurate description of your diet.

Tuna Sandwich
24 g C
19 g P
13.4 g F
293 Calories

Proteins and carbohydrates provide us with about **4 Cal** of energy input per gram of nutrient metabolized, and fat provides about **9 Cal** per gram.

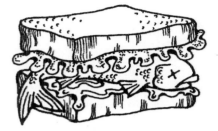

BASIC CALCULATIONS

STEP 1

The Food Composition Tables are used to determine the weight (in grams) of each nutrient in the tuna sandwich.

STEP 2

The weight of each nutrient category must be multiplied by the **Caloric Conversion Factor** for that nutrient. Calculate the Caloric value of each nutrient in the tuna sandwich.

Tuna Sandwich		Caloric Conversion Factor		Caloric Value
24 g C	x	4 Cal per gram	⟶	_____
19 g P	x	4 Cal per gram	⟶	_____
13.4 g F	x	9 Cal per gram	⟶	_____
			Total =	_____ Cal

STEP 3

The final step is to calculate the percentage of Calories contributed by each nutrient. *Do the calculations for the tuna sandwich.*

$$\frac{\text{Caloric Value of a Nutrient}}{\text{Total Calories in the Food}} \times 100 = \% \text{ Based on Calories}$$

$$C = \frac{\qquad}{293} \times 100 = \underline{\qquad} \%$$

$$P = \frac{\qquad}{293} \times 100 = \underline{\qquad} \%$$

$$F = \frac{\qquad}{293} \times 100 = \underline{\qquad} \%$$

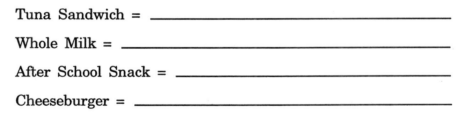

NOW

Let's see if you can calculate the percentage of Calories for each nutrient in the entrées you analyzed in Activity #2.

Entrées	Caloric Value	% of Total Calories
Whole Milk C = g x 4 → P = g x 4 → F = g x 9 →		
After School Snack C = g x 4 → P = g x 4 → F = g x 9 →		
Cheeseburger C = g x 4 → P = g x 4 → F = g x 9 →		

? QUESTION

1. Look at your answers to question #5 in Activity #1. How would you describe these foods now?

   Tuna Sandwich = _____

   Whole Milk = _____

   After School Snack = _____

   Cheeseburger = _____

2. Calculate the Caloric percent of fat in a 3 ounce piece of sirloin steak using the following data: P = 20 g; C = trace; and F = 27 g.

   _____ % fat

3. A green leafy salad is considered to be low in fat and low in Calories. What happens to the fat content when you add a tablespoon of salad oil or creamy dressing?

ACTIVITY #4

"DIETING STRATEGIES"

There are many different ways of cutting Calories out of the diet. We will analyze three common dieting strategies.

EAT ALL YOU WANT OF THIS OR THAT

There are *two* variations of this approach to dieting. One method says that you must eat a certain amount of a particular food *before* eating your meals. For example, the **Eat-Six-Grapefruits-Per-Day** diet instructs you to eat two grapefruits before each of the three meals of the day.

The second variation of this diet approach says that you can eat all you want from a narrow list of foods. For example, the **Eat-All-of-the-Hard-Boiled-Eggs-and-Oranges-You-Want** diet.

NOW

Refer to the Food Composition Table on the last page of this lab.

Let's analyze the Caloric content of the *Eat-All-You-Want-of-This-or-That* diet. Use the Food Composition Table to determine the number of Calories in:

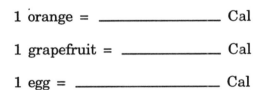

1 orange = _____ Cal

1 grapefruit = _____ Cal

1 egg = _____ Cal

? QUESTION

1. Assume that a person is in energy balance if they keep their food intake to 1600 Calories per day. How many total Calories are in six grapefruits? _____ Cal

2. How much food (compared to normal) are you likely to eat after having two grapefruits before each of your three meals? Your answer should reveal the "trick" of how this diet works.

3. Let's assume that most people on the second variation of this diet wouldn't eat more than 3 eggs and 2 oranges for each of the three meals of the day.

9 eggs = _____ Cal
6 oranges = _____ Cal

The *Eggs-and-Oranges* diet is how many Calories less than the 1600-Cal energy-balance diet?

4. It takes about a 4200-Calorie deficit to lose a pound of body fat. How many days do you have to stay on the *Eggs-and-Oranges* diet until you lose a pound of fat? _____ days.

SPECIAL PREPARED MEALS

This approach to dieting works reasonably well while you are on the diet, and its success is the result of *two* factors. If you are paying for special prepared foods, then you are motivated to stay on the diet longer and not violate it. Also, someone other than you has planned a reasonably good-tasting meal that is not high in Calories.

? QUESTION

1. Why do most people hesitate in using this type of diet?

2. What is the weakness of this approach to dieting after you have stopped the diet?

CUT SOMETHING OUT

There are *two* variations of this approach to dieting. One version is subtle and is illustrated by the ***Butter-and-Jam-on-Toast*** example. The second version is an extreme approach exemplified by the ***Pritikin-type*** of diet during which you eliminate most fat-containing foods.

BUTTER AND JAM ON TOAST

We start with a person who has a well established habit of eating consistent meals. However, recently he has noticed that he is 4 pounds heavier than two years ago. Each morning he eats the same breakfast.

1 bowl of oatmeal
1 piece of toast
1 pat of butter
1 Tbsp. of jam

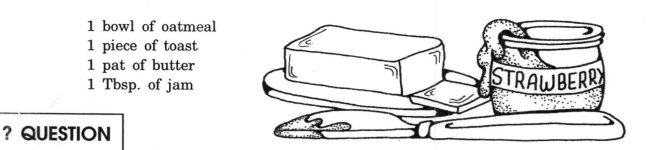

? QUESTION

1. Use the Food Composition Table to determine the Caloric value of the butter and jam.

   1 pat of butter = _____ Cal
   1 Tbsp. of jam = _____ Cal

2. If only the jam were eliminated from his toast each morning, how many days would he be on the diet until he lost one pound of fat? _____ days

3. If both the butter and jam were eliminated from his toast each morning, how many days until he would lose one pound? _____ days

NO MILK OR STEAK

The second version of cutting something out of the diet is illustrated by the next case. Again, we start with someone who has a consistent diet (this keeps the example simple). This person normally eats a 6-oz piece of sirloin steak and one 8-oz glass of milk for each lunch. He has been instructed to replace his normal lunch with a low-fat 400-Calorie pasta and vegetable meal.

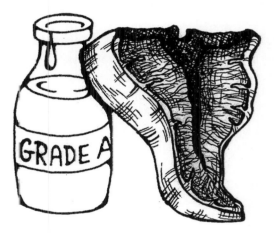

? QUESTION

1. Refer to Activity #3 for the Caloric value of:

   8 oz of whole milk = _____ Cal
   6 oz of sirloin steak = _____ Cal

2. How much Caloric reduction resulted from the switch from the *steak and milk* meal to the low-fat pasta and vegetable meal? _____ Cal per day

3. How many days are required to lose one pound of fat on this low-fat lunch diet? _____

4. What are the advantages of using the *Cut-Something-Out* diet compared to using the other two types of diets?

ACTIVITY #5

"ENERGY EXPENDITURE DURING DAILY ACTIVITIES"

The accurate measure of a person's *energy expenditure* during each of their daily activities is a technical challenge for exercise physiologists. Detailed energy analyses of competitive athletes during performance training have become very imporant for coaches and trainers. Other studies have established estimates for the average person during typical daily activities. Using the estimates, you can predict what will happen to the body energy balance when changes in your job or lifestyle occur. In addition, you can estimate how much more activity is required for you to lose a certain amount of weight.

NOW
---

1. List your general activities during a typical 24-hour period. Record these activities in the Energy Expenditure Table below.

2. Use the information in the **Caloric Conversions for Activities** table on the next page to complete your calculations for energy expenditure. Match each of your daily activities with one listed in the conversion table that most closely approximates the effort expended during your activity. *Be sure to multiply the Caloric Conversion by your weight in kilograms.*

ENERGY EXPENDITURE TABLE

Daily Activities	Caloric Conversion x Your Weight	# of Hours Spent Doing Activity per 24-Hour Day	Calories Expended Doing Activity per Day

Total Calories Expended in 24 Hours = _____

CALORIC CONVERSIONS FOR ACTIVITIES *

Activity	Energy Cal/Hr/Kg
Sleeping	0.9
Lying Still	1.0
Sitting	1.4
Standing, Reading, Writing	1.5
Driving a Car	1.9
Light Exercise	2.4
Walking (3 miles per hour)	3.0
Carpentry, Metalwork, etc	3.4
Bicycling (moderate speed)	3.5
Moderate Aerobic Exercise	4.1
Fast Aerobic Exercise	6.4
Slow Running (5 miles per hour)	8.1
Swimming (2 miles per hour)	8.9
Speed Walking (7 miles per hour)	9.6
Walking Up Stairs	15.7
Rowing in a Race	17.0

* These are approximations based on exercise books that you are likely to find in a library.

? QUESTION

1. Pick one of your usual daily activities. Calculate how many hours of that activity are required to burn 500 Calories per day.

   Description of Activity: _____

   Caloric Conversion for Activity:

   _____ Cal/hr/kg x Your Weight _____ = _____ Cal/Hr

   Hours of the Activity to Burn 500 Calories Per Day = _____

2. A 12-oz beer has about 170 Calories and a 12-oz cola has about 130 Calories. How many minutes of the activity in question #1 are required to burn off a can of cola or beer?

   12 oz cola = _____ minutes of activity

   12 oz beer = _____ minutes of activity

ACTIVITY #6

"SPECIAL CONSIDERATIONS"

Scientific investigations have revealed much about diet analysis and energy balance in the human body. Few of these insights are discussed in popular magazine articles. Yet, some of the discoveries offer possible explanations for the differences in weight gain among people. The extra energy output by **foot bouncing** and by **brown fat** is investigated in this Activity. You will also discover some unexpected **effects of exercise on appetite**. These "special considerations" are only a few of the ideas about diet and exercise that are available to you in scientific books.

FOOT BOUNCING

Either you or someone you know is a foot bouncer. He or she just can't sit still. The foot bounces or the legs move back and forth. Two interesting questions about these people have been tested: (1) Do they burn more Calories than people who sit still? and (2) Do they move more after a big meal? The answer to both of these questions is yes, and we can estimate how much extra energy is expended by their "foot bouncing."

NOW

1. Estimate the number of hours that you spend sitting during a typical day. Include sitting in class, watching TV, studying, talking with others, etc. _____ hours per day

2. Look at the Caloric Conversions for Activities table in Activity #5, and determine the Caloric *difference* between "sitting" and "driving a car." We will assume that the extra movements while driving a car are equivalent to the "foot bouncing" movements.

   _____ Cal/hr/kg extra for "foot bouncing"

3. Calculate the daily Caloric expenditure for "foot bouncing" by multiplying the number of hours spent sitting x the caloric conversion x your weight. _____ Cal

4. Assume that the average "non-foot-bouncer" person burns 1600 Cal per day. What percent more Calories is being burned by the foot bouncer? _____ %

BROWN FAT

Some of the fat cells in your body are specialized to help maintain body temperature. These special **brown fat** cells burn fat and produce extra heat. This allows you to keep warm on a cold day. Some people give off much more heat than others after eating a big meal. Their brown fat may be burning the extra fuel from a big meal. Therefore, some of those extra Calories won't be stored as fat. The heat produced by these people suggests that their metabolic rate is at least 10% higher than normal. In contrast, people with less brown fat may be storing the extra food energy as fat instead of metabolizing it.

? QUESTION

1. Find the energy expenditure for "sitting" in the Caloric Conversions for Activities table. Use this estimate for the normal basic metabolic rate. If "brown fat" increases the energy output to 10% above normal, what is 10% of the sitting energy? _____ Cal/hr/kg

2. If this extra heat-production effect lasted for 10 hours in a 60-kg person, then how many extra Calories would the "high-brown-fat" person burn in a day? _____ extra Calories

APPETITE AND PHYSICAL ACTIVITY

Scientific studies have revealed many factors influencing appetite. For example, if blood sugar or body temperature drops below a threshold level, you get hungry. The amount of chewing during a meal and the stretching of the stomach also reduce appetite. And, of course, we know that early childhood training and daily habits play a role in our desire to eat.

Scientific investigation discovered an unexpected effect of exercise on appetite. It was known that when people worked harder they ate more food. But no one thought to study sedentary people who began a light-activity schedule. The surprising discovery was that the food intake (appetite) actually *decreased*. The implications of these findings are important! Consider the conclusion in reverse: *A person who stops light activity and then becomes sedentary will increase their food intake.* This explains why it is so easy to gain weight in our modern, low-activity lifestyle.

The summary graph on the next page provides many useful insights concerning the effects of physical activity on appetite. Refer to this graph when answering the questions.

BODY WEIGHT AND CALORIC INTAKE
AS FUNCTIONS OF PHYSICAL ACTIVITY

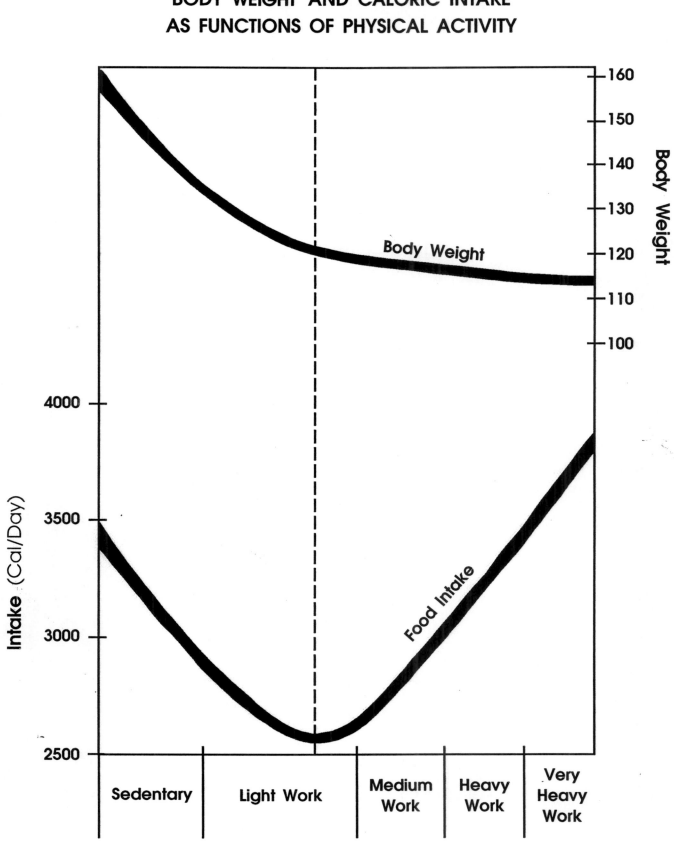

1. How much appetite reduction (in Caloric intake per day) occurred when the study subjects shifted from sedentary to light work? _____ Cal reduction per day

2. *Remember:* A person who shifts from sedentary to light work is also burning more Calories because of the *increased activity*. What is the difference between the sitting and the light-work Caloric expenditures? (Refer to Activity #5.)

   Difference between sitting and light work = _____ Cal/hr/kg

   Your weight = _____ kg

   Total extra Calories burned per day = _____ Cal
   (Assume you do the light work for 8 hours)

3. How many days will it take you to lose 10 pounds? (Consider both the reduced food intake and the increased physical activity.) _____ days

4. The graph shows that the people tested lost about 35 pounds when they shifted from sedentary to light work. The time for this weight reduction was between 4 and 6 months. Do your calculations, based on what you learned in this Activity, generally agree with the study findings? (Multiply the number of days it would take you to lose 10 pounds (question #3) times 3.5 to determine how long it would take you to lose 35 pounds.)

   _____ days to lose 35 pounds

CONCLUSION

We hope that you have gotten value from the scientific approach to diet analysis as presented in this lab. Our discussions are only a beginning of many considerations about diet that have been discovered. There are important questions remaining to be answered. Visit your library for more information. And by the way, have you ever considered a career in nutritional physiology?

FOOD COMPOSITION TABLE

FOOD	C (grams)	P (grams)	F (grams)
Bread (1 slice)	12	2	0.7
Butter (1 pat)	trace	trace	4
Catsup (1 tablespoon)	4	trace	trace
Cheese (1 ounce)	1	7	9
Cracker (1 saltine)	3	0.4	0.5
Egg (1 medium)	trace	6	6
Grapefruit (1 medium)	24	2	trace
$\frac{1}{4}$ pound Hamburger Meat	0	28	23
Jam (1 tablespoon)	14	trace	trace
Lettuce (1 leaf)	trace	trace	trace
Mayonnaise (1 tablespoon)	trace	trace	11
Orange (1 medium)	16	1	trace
Peanuts (1 cup)	27	37	65
Salad Greens (1 cup—no dressing)	7	2	trace
Salad Oil (1 tablespoon)	0	0	14
Sirloin Steak (3 ounces)	trace	20	27
Tuna (2 ounces)	0	15	1
Whole Milk (8 ounces)	12	9	9

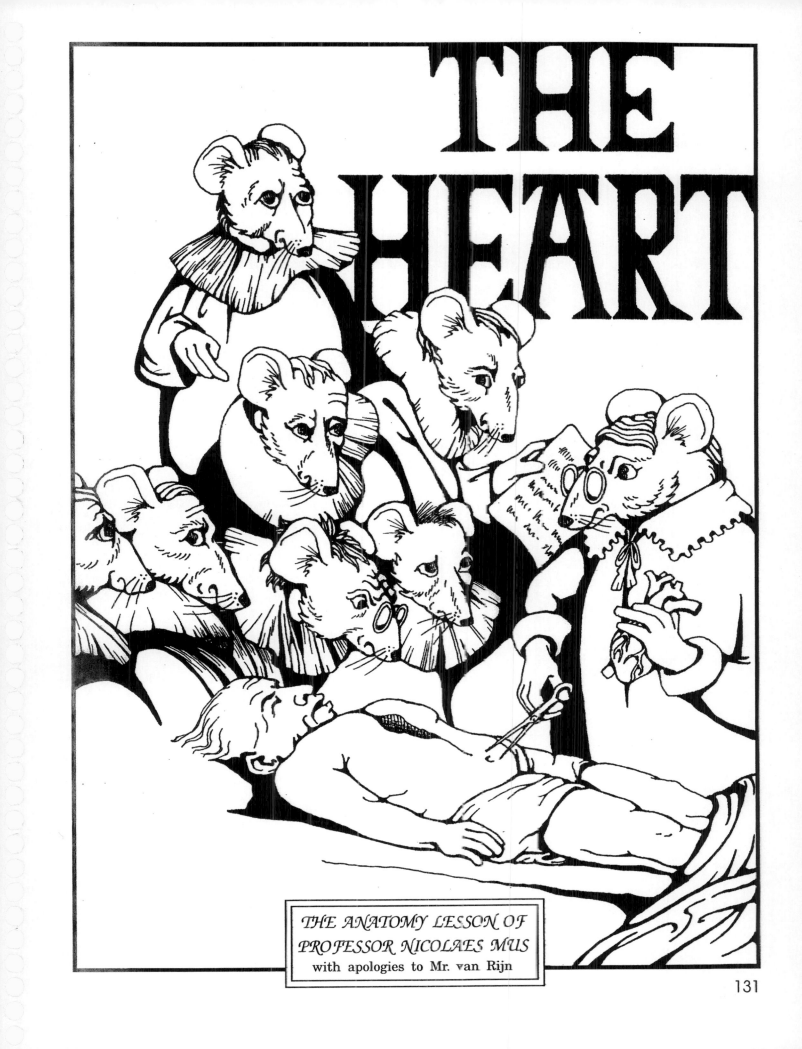

THE HEART

THE ANATOMY LESSON OF
PROFESSOR NICOLAES MUS
with apologies to Mr. van Rijn

Summary Questions

1. Describe the circulation of blood through the heart including how the blood reaches the lungs and the rest of the body.

2. What structures ensure that blood flows only in one direction through the heart? Explain how they operate.

3. Heart sounds are most closely associated with . . .

4. On average, if a person has a heart rate that is elevated 20% beyond normal, about how much reduction in longevity would be expected?

5. Name and describe the two blood pressure readings measured using a stethoscope.

6. What is collateral circulation?

7. Discuss the role of cholesterol, arteriosclerosis, and clots on circulation.

8. What function is being recorded in each part of an EKG: P wave, QRS wave, and T wave?

THE HEART

INTRODUCTION

The heart is an incredible pump. It can beat 2.5 billion times in a lifetime, pumping 35 million gallons of blood. Furthermore, the heart is capable of varying its output between 50 ml and 250 ml per stroke. It can contract at rates from 60 to 160 beats per minute. You can't buy a better pump!

This week's lab will include some aspects of heart structure and function. Also, you will learn to measure blood pressure, and discover the medical implications related to heart function.

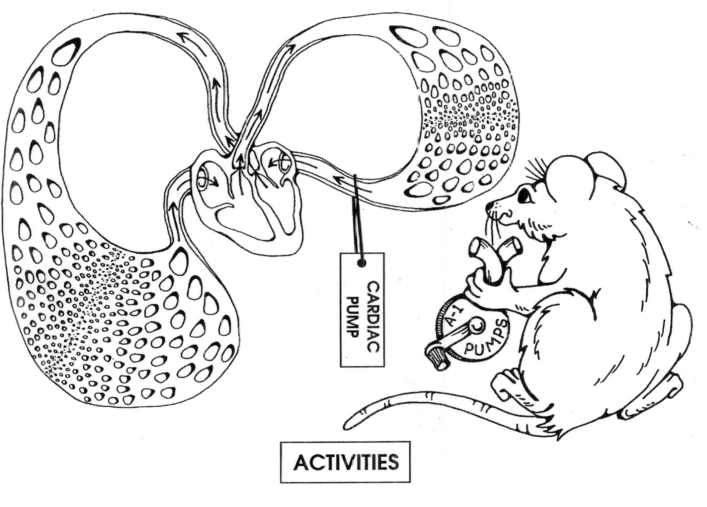

ACTIVITIES

ACTIVITY #1

"THE HEART AS A PUMP"

When you are resting, your heart pumps about 5 liters of blood per minute. This is about the same rate as a slow flow of water from the bathroom faucet when you brush your teeth. During strenuous exercise your heart can pump 30 liters of blood per minute. This is about equal to the water flow when you fast-fill the bathtub. The heart is capable of this wide range of performance because of its structure.

TWO PUMPS IN ONE

The heart is actually two pumps—a *right pump* and a *left pump*. The right pump delivers blood to the lungs; this route is called the *pulmonary circuit*. The left pump pushes blood to the rest of the body; this route is called the *systemic circuit*.

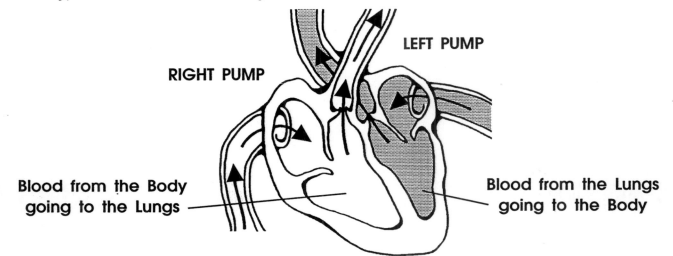

Notice that this diagram is drawn as if the heart is facing you. This means that the right side of the heart is on the *left* side of the drawing. All anatomy diagrams are drawn in this view. Remember this whenever you look at a medical picture.

TWO CHAMBERS PER PUMP

There are *two* chambers in each of the two heart pumps. The top one is a temporary storage chamber called the *atrium*, and the bottom one is a pumping chamber called the *ventricle*. Blood from the body tissues flows into the atrium of the right heart pump. This blood is then pushed through a valve and enters the right ventricle. The ventricle does the hard work of pumping blood out of the heart. The right ventricle pumps blood to the lungs where it is *oxygenated*. While the ventricle is pumping blood out of the heart, the atrium fills with blood entering the heart. This efficient design allows the atrium to quickly refill the emptied ventricle, resulting in a fast-pumping heart.

Oxygenated blood from the lungs enters the atrium of the left heart pump. This blood is then moved into the left ventricle, which pumps the blood to all of the body tissues.

1. Which chamber has to do the most work? (circle your choice)

   Atrium   or   Ventricle

2. Which chamber would have a thicker muscle wall? (circle your choice)

   Atrium   or   Ventricle

3. Which pump has to do the most work? (circle your choice)

   Right Ventricle   or   Left Ventricle

4. Which pump would have a thicker muscle wall? (circle your choice)

   Right Ventricle   or   Left Ventricle

5. The right heart pump moves blood to the _____.

6. The left heart pump moves blood to the _____.

HEART VALVES

Four heart valves are strategically located to prevent backflow as blood moves through the heart. These valves are like one-way doors—they only open in one direction. There is a *chamber valve* between each atrium and ventricle. These two chamber valves ensure that blood will not flow back into the atria when the ventricles contract.

Blood is pushed out of the ventricles and into the two big arteries leaving the heart. There is a valve in each of these arteries. The *artery valves* prevent backflow into the ventricles once blood has been pumped into the arteries. The four heart valves ensure that blood moves in only one direction through the heart circuit. Each of the heart valves has its own special name, but we'll leave those details to an anatomy class.

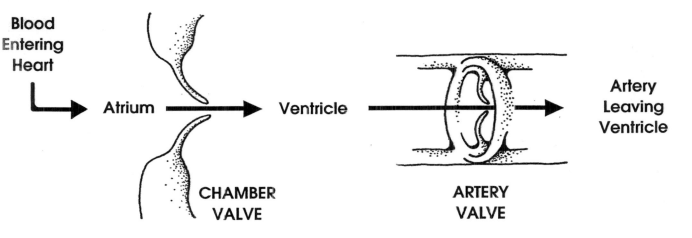

Blood Entering Heart → Atrium → Ventricle → Artery Leaving Ventricle

CHAMBER VALVE   ARTERY VALVE

A heart valve is designed to plug an opening when blood moves in the wrong direction. Think of a valve as being something like a parachute that is attached to the heart or artery wall. If blood moves in the wrong direction, the "parachute" (valve) fills with blood and expands to plug the opening. When the blood moves in the correct direction, the valve collapses like an upside-down parachute. This allows the blood to easily pass through the valve.

EXAMPLE

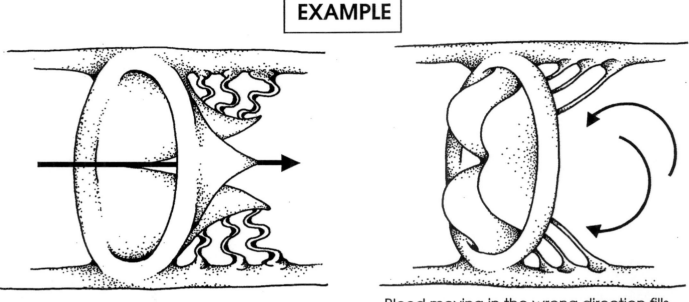

Blood moving in the correct direction pushes the valve aside, and blood enters the ventricle.

Blood moving in the wrong direction fills the valve which plugs the opening so that blood cannot re-enter the atrium.

The heart valves are *flexible* so that they can fill with blood as shown above. There are *special cords* attaching the valve to the heart wall. These cords operate similar to the ropes of a parachute.

? QUESTION

1. What would happen to blood flow if one of the valve "cords" broke? Be specific.

2. What would happen to blood flow if one of the valves was scarred by disease, narrowing the opening?

3. If you had a moderate heart valve problem, what would the heart have to do to compensate?

4. On which side of the heart would a moderate heart valve problem have more consequence to your health? Explain your answer.

GO GET

A sectioned sheep heart from the display table.

NOW

1. See if you can identify the four chambers of the heart. *Remember:* One of the ventricles should have a thicker muscle wall.

2. Find a heart valve and feel the valve to determine its flexibility. Can you find the valve "cords"?

3. Show your instructor when you can identify all of these structures.

4. Draw a simple sketch of the dissected heart. This will remind you of what you saw in case you are tested on it later.

Dissected Heart

ACTIVITY #2

"HEART SOUNDS"

The heart sound is often described as "lub-dub." You might think that the two parts of this sound come from the separate contractions of the upper and lower chambers of the heart. That is not correct. Actually, these sounds are more closely associated with the closing of the heart valves.

The first sound, **lub**, happens when the blood vibrates after the **chamber valves close** between the atria and ventricles. The second sound, **dub**, occurs when the blood vibrates just after the heart **artery valves close**. The sounds are created by vibration waves. With the aid of a stethoscope, a physician can hear these heart sounds and determine if there has been any damage to the valves.

GO GET	

A stethoscope.

NOW

1. Clean the earpieces of the stethoscope with a cotton ball soaked in alcohol. *Always repeat this procedure whenever another person uses the stethoscope.*

2. Fit the stethoscope earpieces in your ears so that they are comfortable and point slightly forward in the ear passage. (Your ear passage points forward before it turns inwards to the ear drum.)

3. Move the bell of the stethoscope around the *left* side of your chest starting at the lower center notch of your rib cage.

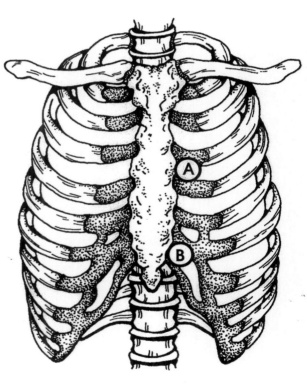

As the stethoscope is moved around the heart area, you will hear the "lub" sound better at some places and the "dub" sound better at other places. An experienced physician or nurse can position the stethoscope to hear each heart valve and determine whether there is an abnormal sound. Abnormal sounds indicate possible valve damage or other circulation problems.

This part gets a little tricky, so read carefully.

Most people expect that the two big heart arteries would exit from the bottom of the ventricles, but they don't. These arteries come out of the top of the ventricles, and arch upwards above the heart. Refer to the diagram under "Two Pumps in One" in Activity #1, and notice the location of the two heart arteries. The artery valve allowing blood flow out of the ventricle is next to the chamber valve controlling flow from the atrium into the ventricle.

The **"dub"** sound of the heartbeat resonates *upward*, which is the direction that the heart arteries leave the heart. In which position of the stethoscope (A or B) do you hear mainly a "dub" sound? _____ This would indicate that you are located near the top of the heart where the heart artery valves make their sounds.

Continue to move the stethoscope to *both sides* of this position until the heart sounds quiet. You are mapping the general location of the top of the heart. Mark the extent of this "dub" zone on the rib cage diagram. The sound zone is bigger than the heart because the sound spreads outward.

The **"lub"** sound of the heartbeat resonates *downward* from the chamber valves, so you hear it best at the bottom of the heart. In which position of the stethoscope (A or B) do you hear mainly a "lub" sound? _____ This would indicate that you are located nearer the bottom of the heart where the sound of chamber valves is loudest.

Continue to move the stethoscope to the *left* of this position until the loud "lub" sound quiets. You are mapping the general location of the bottom of the heart. Mark the extent of this "lub" zone on the rib cage diagram.

FINALLY

Draw an arrow from the center of the "dub" zone to the center of the "lub" zone on the rib cage diagram. That arrow is the general orientation of the heart. The orientation is (circle your choice)

        vertical (straight up and down).

        to the left of vertical (bottom of the heart points to the left).

        to the right of vertical (bottom of the heart points to the right).

This sound technique of inferring the orientation of the heart is less accurate than the EKG method used in hospitals. The EKG can reveal whether the heart has enlarged on its left or right side. Enlargement is an indication of a heart or circulation abnormality.

ACTIVITY #3

"HEART RATE AND THE PULSE"

The heart of a resting person contracts somewhere between 60 and 80 times per minute. These contractions can be counted with the aid of a stethoscope, which you will do in a few minutes. Heart rate can also be determined by counting the number of pulse waves that pass by a spot on an artery. Counting these waves by touch is how you measure the *pulse*.

The *aorta* is the large artery that supplies the blood to all other arteries that feed the body tissues.

Each heart contraction forces a volume of blood (50–250 ml) into the aorta. First, the aorta is "ballooned" out by blood, then the elastic artery wall snaps back an instant later.

The recoil causes the adjacent area of the aorta to balloon out and snap back. The alternate expansion and recoil of the aorta wall "pulses" outward from the heart to the other arteries of the body.

You feel these waves passing by whenever you press a finger on an artery.

The *carotid pulse* is felt when you press your fingers against the side of your throat. The *radial pulse* is felt when you press your fingers on the thumb side of your upward-turned wrist.

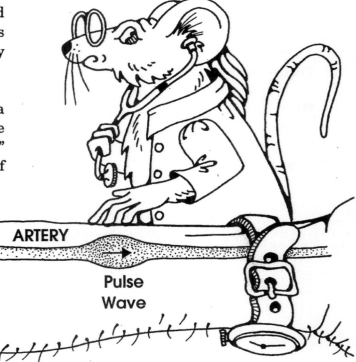

NOW

1. Use the stethoscope to count your heartbeats for 30 seconds.

    Stethoscope Heart Rate = _____ beats per minute

2. Count both your radial pulse waves and carotid pulse waves for 30 seconds.

    Radial Pulse = _____ waves per minute

    Carotid Pulse = _____ waves per minute

140

1. Which had the stronger pulse waves? (circle your choice)

   Carotid   or   Radial

2. Which is closer to the heart? (circle your choice)

   Carotid   or   Radial

3. Which would produce a stronger pulse wave? (circle your choice)

   A smaller heart   or   A bigger heart

4. Arteriosclerosis hardens the artery wall with scar tissue. If arteries have been partially injured by arteriosclerosis and the arterial wall is less flexible than normal, then would the pulse be stronger or weaker than normal?

5. If a person has arteriosclerosis, what happens to the blood pressure near the end of the arteries? (circle your choice) **Hint:** Is some of the energy of the heart contraction "used up" by the pulse wave?

   a. it is the same as normal

   b. it is higher than normal

   c. it is lower than normal

   What would be the consequences?

6. Half of the people have a smaller-than-average sized heart and half have a larger-than-average sized heart. In which group would you expect the *heart rate* to be higher? Explain your answer.

7. In general, females have a higher heart rate than males. What explanation can you give for this difference?

HEART RATE AND LONGEVITY

Research in comparative physiology suggests that the average mammal heart beats about 1.5 billion times before it wears out. Although there are exceptions to this heart longevity rule, it seems to be generally true whether you're a mouse or an elephant. A mouse's heart beats about 10 times faster than an elephant's heart, and a mouse lives about $\frac{1}{10}$ as long.

Based on the mammalian average for total heart beats, the modern human species is predicted to live about 35 years. Humans score above most other species for longevity. This is probably because we are smarter and can avoid more hardships than the average mammal. However, we also have a limit—somewhere around 2.5 billion beats—if we are lucky enough to survive disaster and illness. How you spend these heartbeats is partly determined by the activities in your lifestyle.

Let's assume that the longevity rule is generally true for humans. A woman who is already doing enough daily activity to keep her heart healthy, asks the question, *"If I train in a very strenuous sport for 4 hours a day beyond my normal activity, then how much am I shortening my life by doing this sport?"* Assume that her normal heart rate of 70 is elevated to 120 during the heavy training.

? QUESTION

1. How many *extra* heartbeats does she use per day of sport training? _____

2. Her normal heart rate of 70 per minute means that she would use 100,800 heart beats on a normal day without sport. If an extra 100,800 beats shortens her life by one day, then how many days of sport training does it take to shorten her life by one day? _____

3. How many years of sport would shorten her life by one year? _____

4. Let's assume that this person is considering 8 hours of strenuous sport per day. How many years of this sport activity would it take to shorten her life by one year? _____

   Before we ascribe too much importance to a higher heart rate, remember that females generally live 10% longer than males even though females have a 10% higher resting heart rate. Obviously, other important factors affect longevity.

5. Smoking elevates the heart rate about 10% above normal; so does drinking 2 to 4 cups of coffee per day. If you were a smoker or a coffee drinker for 40 years, how many years of longevity might be lost due to the increased heart rate alone (not taking into account the obvious health risks of tobacco and caffeine)? _____

6. Negative stress can elevate the heart rate 10–20% above normal. How many lost years of longevity might result from a 20-year stress-filled job that elevated heart rate 20% above normal?

ACTIVITY #4

"HOW TO MEASURE BLOOD PRESSURE"

Knowing how to measure your blood pressure is one of the best health-maintenance tools you can have. We offer this Activity to promote your good health.

BLOOD PRESSURE

Blood pressure in body arteries is created by the contraction of the left ventricle. As you would expect, the pressure is highest when the chamber contracts. Pressure during heart contraction is called the *systolic pressure*. When the ventricle relaxes the blood pressure drops. However, instead of dropping to zero, the blood pressure is partially maintained by the recoil of artery walls that are stretched by blood pumped out of the heart. The lower artery pressure during the relaxation of the ventricle is called the *diastolic pressure*.

Medical books state that the typical resting blood pressure is 120 over 80. The 120 refers to the systolic pressure and the 80 refers to the diastolic pressure.

MEASURING BLOOD PRESSURE

The method we will use to measure blood pressure is fairly simple. It is easier to show you how to measure blood pressure than to explain all of the details in writing. Therefore, most of the instructions will come from your lab teacher.

Read through the general steps of the procedure and the hints that follow before you begin practicing.

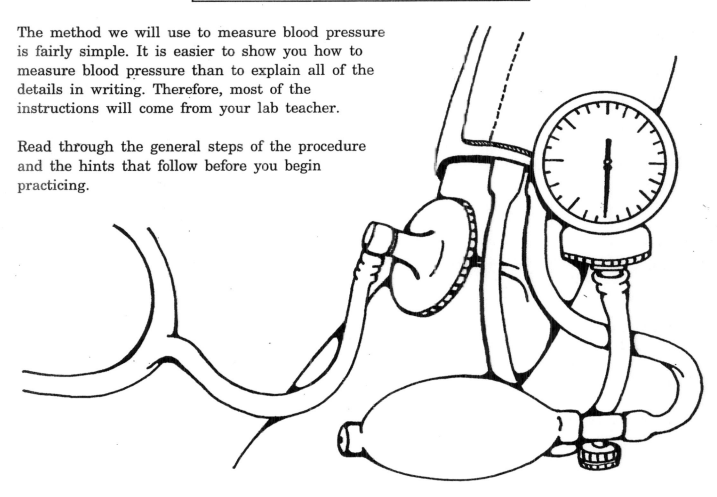

A blood pressure cuff.

STEP 1
Fasten a pressure cuff around your upper arm. Place the stethoscope diaphragm over the brachial artery (*inside bend of elbow*). Pump the cuff full of air until all blood is stopped in the brachial artery. The thumping heart sound that you hear through the stethoscope will fade and disappear as the cuff pressure is pumped above the systolic pressure.

STEP 2
Release the pressure on the cuff by slowly opening the valve. Listen to the brachial artery with the stethoscope.

STEP 3
When you first hear a "thumping" sound, read the gauge on the pressure cuff. This is the *systolic pressure*. The blood is just starting to squirt past the cuff during the contraction of the heart.

STEP 4
Continue to release the pressure on the cuff until the thumping sound disappears. Read the pressure gauge. This is the *diastolic pressure*. The blood flows past the cuff during both the contraction and relaxation of the ventricle. The sound disappears when the flow changes from a pulsating squirt to a constant flow.

HINTS

1. Don't pump the pressure cuff over 150 until you've practiced the technique several times.

2. Don't keep pressure on your arm for more than 30 seconds.

3. Let your arm rest for at least 2 minutes after each reading before taking another measurement. This is especially important while you are learning the technique.

4. Take your time. Learn this procedure well. It is important that you can measure your own blood pressure. Checking your blood pressure every few months provides you with a thorough understanding of your normal physiology. When an abnormal change occurs, you can seek medical advice.

NOW

Record your blood pressure while sitting: _____ B.P.

With the blood pressure cuff still on your arm but not pumped up, run in place for 30 seconds. Measure your blood pressure after exercise: _____ B.P.

COLD WATER TEST

This test is used to determine the effect of a sensory stimulus (cold) on blood pressure. The normal reflex response to a cold stimulus is a slight increase in blood pressure (both systolic and diastolic). In a normal individual, the systolic pressure will rise no more than 10 mm Hg, but the increase in a **hyper-reactive** individual may be 30 to 40 mm Hg. We will discuss the implications of these responses later.

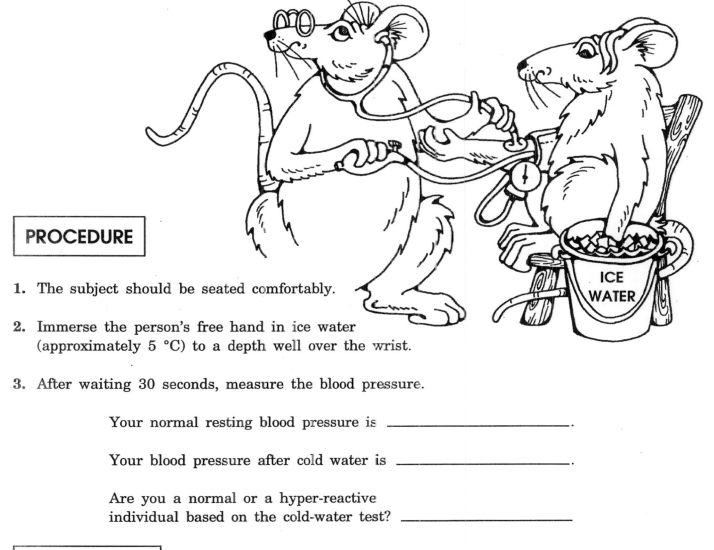

PROCEDURE

1. The subject should be seated comfortably.

2. Immerse the person's free hand in ice water (approximately 5 °C) to a depth well over the wrist.

3. After waiting 30 seconds, measure the blood pressure.

   Your normal resting blood pressure is _____.

   Your blood pressure after cold water is _____.

   Are you a normal or a hyper-reactive
   individual based on the cold-water test? _____

IMPLICATIONS

There is some evidence that people showing a hyper-reaction to a cold stimulus may have a greater chance of developing high blood pressure later in life. Perhaps there is some minor defect in the physiology of these people. Or they may have inherited a more reactive nervous system—favored in hunter-gatherer times—but more easily overstimulated by our modern chaotic life. We just don't know.

Before giving too much importance to the results of the cold water test, remember that lack of exercise, high salt intake, unhealthy diet, and stressful situations are known causes of high blood pressure, and are mostly under your control.

ACTIVITY #5

"MEDICAL IMPLICATIONS"

There are many medical implications related to what you have learned about the heart, and several of these health issues are discussed in this Activity. *Remember:* You should always ask your physician to explain health-related problems in a way that you can easily understand. And go to the library. Inform yourself.

ABNORMAL HOLES IN THE HEART

The heart in a fetus has a hole between the right and left atria. This opening allows fetal blood to partially bypass the lung circuit since the lungs aren't needed to get oxygen during life in the womb. Normally, the atrial hole closes shortly after birth. A birth defect results if this hole does not grow closed. Another birth defect occurs when there is an abnormal opening between the right and left ventricles. Both of these heart abnormalities can have serious health implications if left uncorrected.

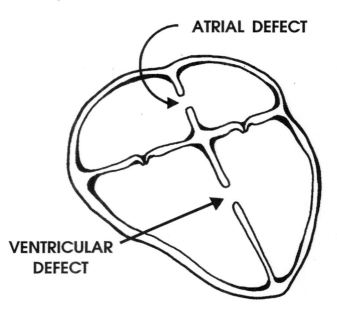

ATRIAL DEFECT

VENTRICULAR DEFECT

? QUESTION

1. What important molecule is carried by blood entering the left heart pump and is not in the blood entering the right heart pump?

2. If the fetal hole between the right and left atria does not close, what happens to the blood in these two chambers?

3. Which ventricle operates under the most pressure (does the most work)?

4. What would happen to the pressure in the two ventricles if there was a hole between them?

5. What would the heart do to compensate for the pressure problem created by a ventricle hole? *Hint:* People with this abnormal hole must have it repaired while they are young or they won't live long.

CORONARY ARTERIES

The heart muscle works very hard and it must be supplied with oxygen and nutrients just like any other part of the body. The vessels that supply blood to the heart muscle are called the *coronary arteries*.

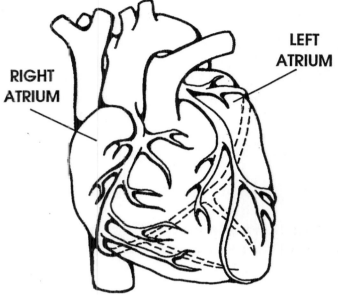

There are two important circulation patterns that you can see in this diagram. The right atrium is fed only by the right coronary artery, and the left atrium is supplied only by the left coronary artery. However, each coronary artery supplies blood to parts of *both* ventricles. The ventricles do more work than the atria, and must be supplied with more blood.

Another important aspect of vessel structure in the coronary arteries is the connection between the arteries. Connections between arteries are called *collateral circulation*. These connections are alternate routes of blood flow to tissue if one path is blocked. Some parts of the heart have no collateral circulation. Other parts have only very small-diameter collateral vessels because they are unused. Also, there are different amounts of collateral vessels among people. Can you find a collateral vessel in the heart diagram above? Color that vessel.

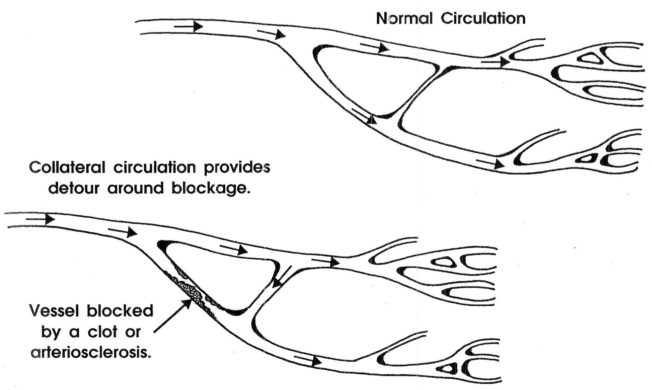

147

1. A patient is told that she has a narrowing of the right coronary artery. Which chambers of her heart are going to be the most affected by this disorder? _____

   Which chamber on the right side has to do the most work and could be the most serious health concern? _____

2. A group of patients were told that they had plugged arteries in their hearts. In addition, they had all suffered a similar size of heart attack. All of these patients survived. Some of them had parts of their injured hearts return almost to normal after several months. The other patients had no such luck. Explain these differences in terms of coronary circulation.

3. A heart attack on which side of the heart would probably cause the most serious immediate risk to the person?

EKG

The *electrocardiogram* (called *EKG* or ECG) is a recording of the small electrical currents produced by the contracting and relaxing heart muscle. These electrical patterns indicate whether there is a normal or abnormal functioning of the heart. A normal EKG is shown here.

The *P wave* is a recording of the electrical activity in the atria, and it is especially important in diagnosing problems in the heart's natural *pacemaker*.

The *QRS wave* is a recording of the activating current in the ventricles. This current travels along a special conducting pathway 10 times faster than it would be transmitted through normal heart muscle. The result is that all the muscle cells in both ventricles are stimulated at the same time, causing these chambers to contract quickly and strongly.

The *T wave* occurs just after the ventricles contract, and is a recording of the normal recovery phase of the ventricles. This is a period when the muscle cells perform various biochemical reactions that prepare the ventricles for the next contraction.

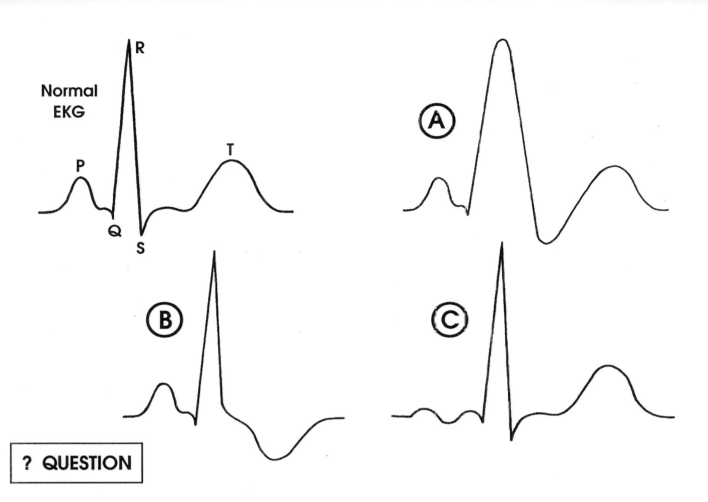

Normal EKG

P Q R S T

A

B

C

? QUESTION

1. A person who drinks a lot of coffee complains that his heart beats irregularly. Which one of the above EKG waves reflects this problem? _____ Explain your answer.

2. A person has a greatly enlarged heart from the overwork created by long term high blood pressure. Which abnormal EKG wave reflects this problem? _____ Explain your answer.

3. A person with poor coronary circulation to the heart muscle has some heart injury, but may not have suffered death of the heart muscle. Which abnormal EKG wave reflects this problem? _____ Explain your answer.

4. Why is a heart attack sometimes called a "coronary"?

151

1. List five specific health improvements that your body makes as a result of a conditioning program.

2. Define oxygen debt.

3. Why would one person favor heart rate and another person favor blood pressure during the recovery period after the Step Test?

4. If a sedentary person begins a moderate exercise program, what will happen to their resting heart rate?

5. Discuss the specific relationship of vessel diameter to blood flow and heart workload.

6. If exercise improves the performance of the heart, why would that increase longevity?

CONDITIONING

INTRODUCTION

The improvement of cardiovascular fitness resulting from daily exercise is called **conditioning**. Scientific investigations have discovered that characteristic body changes occur during a physical training program. For example, your resting heart rate and blood pressure are likely to decrease. Your heart muscle is strengthened, and its own circulation improves. The blood volume and the amount of hemoglobin (the molecule that carries oxygen) found in red blood cells usually increase. Furthermore, the air capacity of your lungs increases. All of these changes are necessary to the performance of a competitive athlete, and are very important to the health of an average person.

The scientific study of conditioning has revealed that specific body changes result from different training programs. We know how to condition muscles for speed of contraction, strength of contraction, and endurance. Each of these muscle adaptations can be developed during a particular type of training. Furthermore, rapid improvement can be achieved by an out-of-condition person (couch potato), whereas, someone who is already moderately fit requires a more intensive program to improve conditioning.

Some discoveries about physical conditioning were unexpected by American sports trainers. They gained these insights from exercise scientists in other countries. For example, optimum performance in certain sports is achieved by those athletes who can train themselves to be aware of their physiological processes (such as heart rate). Competitive success occurs only when they learn to maintain the most efficient heart rate for their physiology during a particular sporting contest.

The good news for us is that the science of conditioning has much to offer the average person. The bad news is that there is much misinformation generated by nonscientific theories, and most of it is targeted at and used by the average person. In this lab you will investigate several effects of exercise on the cardiovascular system. Hopefully, you will become interested enough to read more scientific articles and books on conditioning and training.

ACTIVITIES

153

ACTIVITY #1

"EXERCISE AND RECOVERY"

During exercise, muscle contraction is fueled by a high-energy molecule called **ATP**. This power molecule is produced by the metabolism of food.

OXYGEN DEBT

When your body metabolizes food, oxygen is consumed. Oxygen is replenished by the respiratory and circulatory systems. During light exercise, the demand for oxygen is easily satisfied by your increased breathing and blood flow. However, if you exercise at a very strenuous pace, your body cannot keep up with the oxygen consumption of your muscles. Strenuous activity creates an **oxygen debt** that must be repaid. This requires that you stop and rest.

THE STEP TEST

It takes time to recover from an oxygen debt. The specific changes in your *heart rate* and *blood pressure* during the recovery period will reveal several things about your cardiovascular anatomy and fitness. A simple method of creating an oxygen debt is the **Step Test**.

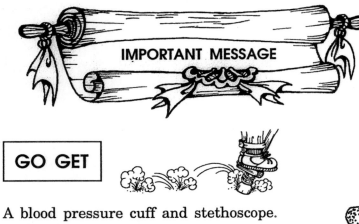

IMPORTANT MESSAGE

Your group must develop an organized plan to accurately measure both heart rate and blood pressure after the Step Test. Read the hints and make a plan.

GO GET

A blood pressure cuff and stethoscope.

HINTS

1. Read all directions in the procedure before you begin. It is absolutely necessary to organize your group so that you can get accurate results.

2. Put the blood pressure cuff on the arm before starting the test. This will allow you to take quick readings during the recovery period.

3. Test one person in the group at a time. Assign one person to count the pulse and another person to measure the blood pressure.

PROCEDURE

1. Everyone in the group must do 5 minutes of light warm-up exercise, such as a quick walk around campus. Do it now!

2. To make sure the heart rate has returned to normal, wait 10 minutes after the warm-up before beginning the Step Test.

3. Record the subject's resting *heart rate* and *blood pressure* before starting the Step Test.

4. The exercise part of the Step Test involves stepping up and down a step at the rate of one step every two seconds (30 steps per minute). One person in the group should count out loud at the proper rate to cue the test subject.

5. Continue stepping until the subject feels tired (1–5 minutes, depending on fitness). The objective is to exercise to a point where you feel that you will need some rest to recover your oxygen debt, but not to total fatigue.

6. Immediately measure the heart rate when the subject stops exercising.

7. Continue recording the subject's heart rate every minute until it slows to the pre-exercise resting rate (#3 above).

8. Immediately measure the subject's blood pressure *when the heart rate slows to the normal resting rate.*

9. If the blood pressure remains above the pre-exercise resting rate, continue to measure it until it returns to normal. Keep track of the time it takes to reach normal.

10. Repeat the procedure for each member in your group.

```
┌─────────────────────────────────────────────────────────────────┐
│                         DATA CHART                                │
├─────────────────────────────────────────────────────────────────┤
│   Resting Blood Pressure: _____                         │
│   Resting Heart Rate: _____                             │
│   Step Test for _____ minutes.                          │
│   Heart Rate immediately after Step Test: _____         │
│   Heart Rate 1 minute after Step Test: _____            │
│                          2 minutes _____                │
│                          3 minutes _____                │
│                          4 minutes _____                │
│                          5 minutes _____                │
│                          6 minutes _____                │
│                          8 minutes _____                │
│                         10 minutes _____                │
│   Blood Pressure when the Heart Rate recovers: _____            │
└─────────────────────────────────────────────────────────────────┘
```

RECOVERY CURVES

Your recovery rate after exercise depends on the severity of the oxygen debt that developed during the Step Test and the fitness of the systems responsible for correcting that debt. Plot your heart rate data on the recovery graph. Indicate when your blood pressure returned to normal.

Recovery curves can be used to evaluate the amount of fitness improvement developed during a physical training program. However, to be accurate, the Step Test must be standardized with regard to the exact amount of stepping done. You will use your data in a very different way. This experiment will reveal whether you have a heart that recovers from oxygen debt by *beating faster* (heart rate) or by *beating harder* (blood pressure).

RECOVERY GRAPH

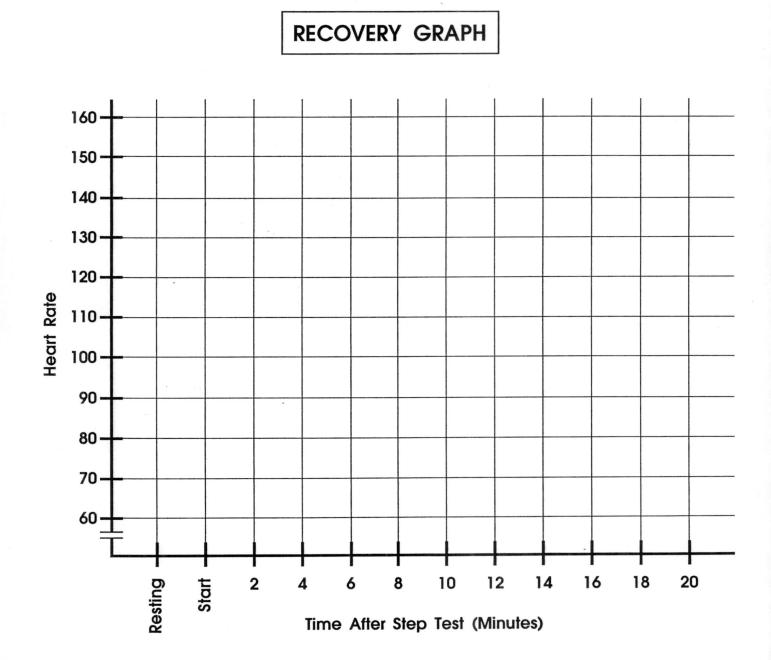

METHOD OF RECOVERY
HEART RATE OR STROKE VOLUME

There are two ways of increasing heart output during exercise. The first is to pump faster (heart rate), and the second is to pump more blood per beat (stroke volume). Both methods allow us to recover from an accumulating oxygen debt. Heart output usually shows an emphasis on one or the other method depending on the nature of that particular heart. For example, people with a larger than average heart have a greater range in stroke volume possible per heart beat. When the person is resting, the heart may pump only 70 ml per beat, but during exercise a healthy large heart may pump 250 ml per beat. The smaller heart does not have this same range in stroke volume, and must handle more of its output demand by increasing the heart rate. Your heart will favor either *stroke volume* or *heart rate,* and this can be determined by the Step Test recovery data.

If your heart quickly returned to the resting heart rate but your blood pressure remained above normal for a longer time, then your heart probably favors **stroke volume**.

If your heart rate was higher for a longer time and the blood pressure was about normal when heart rate recovered, then your heart favors **heart rate** as the method of recovery after exercise.

Stroke Volume **X** Heart Rate = Heart Output

Greater Stroke Volume ➤ Heart Rate can be slower

Smaller Stroke Volume ➤ Heart Rate must be faster

? QUESTION

1. Which took longer to recover during your step test? (circle your choice)

        heart rate   or   blood pressure

2. Which method of recovery does your heart favor? (circle your choice)

        heart rate   or   stroke volume

3. Do you engage in regular exercise?

Remember: If a sedentary person begins a moderate exercise program, then their heart will strengthen and pump more blood per beat. This heart conditioning usually reduces the heart rate during exercise.

ACTIVITY #2

"BLOOD VESSEL HEALTH"

Arteriosclerosis is a common cardiovascular disease in modern people. Although you learn about it in relation to older people, it is a disease that develops over a long period of time. Many factors contribute to this disorder, including stress, sedentary lifestyle, a high-fat diet, and certain genetic tendencies you may have inherited from your parents. The health significance of arteriosclerosis is created by the steady decrease in blood vessel diameter. In this Activity you will examine the *effects of decreasing vessel diameter on blood flow* and what the heart must do to compensate.

BLOOD FLOW AND VESSEL DIAMETER

Your group will measure the flow of water through *three sizes* of tubing (small diameter, 2x-bigger diameter, and 4x-bigger diameter). The results will demonstrate the effects of reduction in vessel diameter (as happens in arteriosclerosis).

GO GET

The vessel diameter flow apparatus.

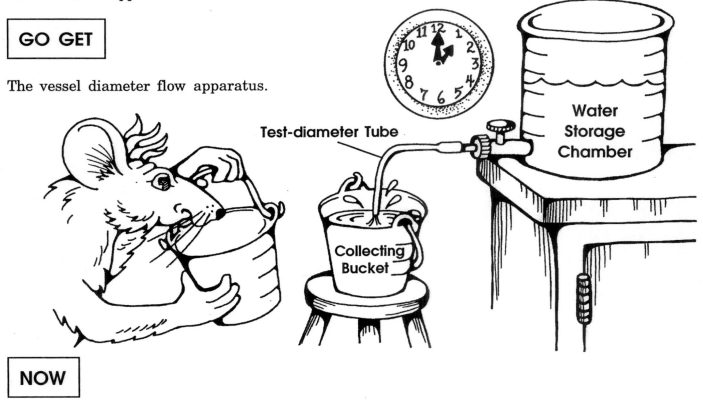

NOW

Read the following directions and modify them to match the setup used in your lab class.

1. Fill the water storage chamber with water to the same level for each test, and prepare to catch the outflowing water with a collecting bucket.

2. Test the smallest-diameter tube first.

3. Open the flow through the test-diameter tube. Collect the water during exactly one minute of open flow. Measure the volume of water in the collecting bucket, and record it.

4. Refill the storage chamber, and repeat the test with the next-size tube: the 2x-bigger diameter. Record the volume during one minute of flow.

5. Refill the storage chamber, and repeat the test for the 4x-diameter tube. The storage chamber may empty before one minute. If so, make a note of the time it took to empty and correct the volume to reflect what the flow would have been in one minute. Ask your instructor to check your logic in making these calculations.

Flow through the smallest-diameter vessel in one minute = _____

Flow through the 2x-diameter vessel in one minute = _____

Flow through the 4x-diameter vessel in one minute = _____

? QUESTION

1. How much water flows through the 2x-diameter vessel compared to the smallest-diameter vessel?

   Hint: Divide $\dfrac{\text{Volume of 2x-diameter}}{\text{Volume of Smallest-diameter}}$ = _____ x more water

2. How much water flows through the 4x-diameter vessel as compared to the 2x-diameter vessel?

   _____ x more water

3. What is the average increase (multiplier) in flow when you double the diameter of a vessel? (average of #1 and #2)

   _____ x more flow

4. To discover the effects of arteriosclerosis on heart workload, you need to consider these flow values in reverse. How many times (multiplier) less blood flows when the blood vessel diameter is halved? Quartered?

   $\frac{1}{2}$ = _____ x less flow $\frac{1}{4}$ = _____ x less flow

INCREASED HEART WORKLOAD

Heart workload is directly proportional to the resistance in blood flow. A decrease in vessel diameter increases **resistance**. This resistance increases the workload on the heart. *The heart workload is directly proportional to the resistance.* In other words, if the resistance is doubled (2x), then the heart workload is doubled (2x) in order to overcome the increased resistance.

↑ Resistance = ↑ Heart Workload

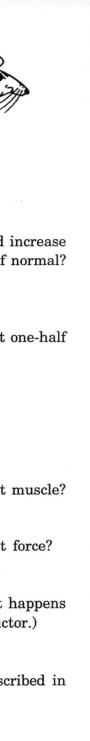

? QUESTION

1. Refer to question #4 on the previous page. How much more does the heart workload increase (what is the multiplier?) when arteriosclerosis reduces vessel diameter to one-half of normal?

2. When you are under stress the nervous system contracts your blood vessels to about one-half their normal diameter. What does this do to your heart workload?

3. What are the two methods of recovery used by the heart to increase its output?

   _____ and _____

4. If a muscle works harder, then what would you expect to happen to the size of that muscle?

5. Arteriosclerosis and stress increase the size of the heart. This is to overcome what force?

6. When the thickness of the heart wall increases in response to high workload, what happens to the inside volume of the heart chambers? (Check your answer with your instructor.)

7. What happens to the heart rate when the heart accommodates to the changes described in question #6?

8. Competitive athletes usually develop larger heart muscles. Why?

9. Do you see any similarities between an older person with arteriosclerosis and an athlete? Explain.

10. Why is a lifetime diet of fried chicken, burgers, cheese nachos, pizza, and ice cream a problem?

Summary Questions

1. What one simple thing can you do to improve your overall health and increase your lifespan? (Not diet or exercise.)

2. What is specific gravity, and what causes it to increase?

3. What is meant by the body's water balance?

4. How does the specific gravity of urine change after drinking a large volume of water? Explain.

5. What does the "% of Water Returned" after the water-loading test tell you about your normal body water level?

6. List the ways that your body gains water, and the ways that it loses water.

KIDNEY FUNCTION

Mammal kidneys are similar to the kidneys of reptiles and birds, but are quite different from those of freshwater fish and amphibians. This clue suggests that our kidneys evolved in response to the scarcity of water in the land environment. Kidneys perform three basic functions. One is to regulate the fluid content in the blood of the organism. Another is to balance the concentration of dissolved salts and other electrolytes (acids and bases) in the blood. While balancing those substances, the kidney must also get rid of the waste products from the foods metabolized by the organism.

During a 24-hour period your kidneys filter about 180 liters of blood. This means that each day your blood is reprocessed over 50 times. If the kidneys fail, it is only a few days before a person is dead. You have heard about the importance of a good diet and regular exercise and a low stress environment. But, has anyone told you that caring for your kidneys is as significant to your health as diet, or exercise, or smoking? What if someone told you that drinking 8 glasses of water a day does more to increase longevity than going to the gym twice a week? Would you believe it?

There is much useful scientific information about the urinary system, but today we only have time to focus on some highlights. In this week's experiment you will investigate the *effects of water-loading on urine concentration and production rates*. Next week you will analyze the results of today's experiments, and review the general features of the urinary system.

ACTIVITIES

ACTIVITY #1

"SPECIFIC GRAVITY OF URINE"

The density of a fluid compared to pure water (S.G. = 1.000) is called the **specific gravity**. If you add salt or any other dissolvable substance to pure water, then the specific gravity of the solution *increases above* 1.000.

The specific gravity measuring equipment used in today's experiment is called a **urinometer**. It responds the same way you do when you float in freshwater or in saltwater. You probably have noticed that it is easier to float in the ocean than in a swimming pool. This is because the ocean has dissolved substances (mostly salt) that increase the specific gravity above 1.000. The greater density of ocean water produces a stronger buoyancy force. The heavier water displaced by your body allows you to float higher in the water.

GO GET

1. A urinometer kit.

2. A sample of 3% saltwater.

3. A sample of distilled water.

NOW

Urinometer Float

Read where the "waterline" of the urine intersects the float.

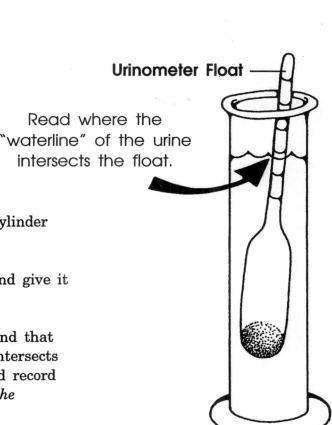

1. Pour the distilled water into the small glass cylinder until it is about 2/3 full.

2. Place the urinometer float into the cylinder, and give it a gentle spin.

3. Notice that the float has a measuring scale, and that the "waterline" of the distilled water sample intersects at some point on the scale. Read the scale and record the specific gravity on the next page. *Repeat the procedure on the saltwater sample.*

SAMPLE	SPECIFIC GRAVITY
Distilled Water	
Saltwater (3%)	

IMPORTANT MESSAGE

Check with your instructor to make sure that you are reading the *decimal position* correctly on the specific gravity scale.

1.000

1.010

1.020

Float Scale

? QUESTION

1. The urinometer _____ in the fluid that has a greater specific gravity. (circle your choice)

    floats higher   or   sinks lower

2. After drinking a lot of coffee or a few beers, perhaps you have noticed that the color of your urine has changed.

    a. Describe what changed about the color.

    b. Do you think there could be a change in the specific gravity? _____
    What gives you that idea?

3. Urinometers are designed to measure the normal range of urine concentration. You observed the position of the urinometer float in 3% saltwater (sea water). Based on your observation, is urine less or more concentrated than sea water?

4. Why can't you use sea water as drinking water?

ACTIVITY #2

"EFFECTS OF WATER-LOADING ON KIDNEY FUNCTION"

Drinking a liter of water should create a **water overload** in your body. During the two hours after drinking the water, you will collect urine samples and record the changes in *urine production* and *concentration*. These changes are produced by the kidneys' readjustment of the body's water balance. The results of this experiment will also reveal whether you have been drinking enough water today or if you are partially dehydrated.

THE BASIC PROCEDURE

The intake of water and the collection of urine samples will be done at home or during lab class, depending on your instructor's approach. All urine samples are measured at the time of production. *Be very careful with the urinometers. They break easily and they cost money to replace.*

The basic procedure is as follows:

1. Drink a liter of water. Then measure your urine volume and its concentration at timed intervals.

2. The volume measurements are fairly easy since all you have to do is pee into the graduated beaker and read the amount. ***Remember:*** Don't pour out the urine until *after* you measure the specific gravity.

3. The specific gravity is measured by placing the urinometer float into a small cylinder partially filled with each sample of urine, and recording where the "waterline" is on the measuring scale. Obviously, the trick is to do all this without spilling urine on yourself or on the floor.

4. Rinse the equipment with plain water between each sample, and you're ready for the next one.

GO GET

1. A urinometer float for each person in your group.

2. A urine collection beaker.

3. A 1-liter drinking container.

THE EXPERIMENT

<table>
<tr><td>STEP 1</td><td>Go to the bathroom and collect a urine sample, and *measure the specific gravity only*. We don't care about the volume of this first reference sample, but you must record your starting specific gravity value in the Data Table.</td></tr>
<tr><td>STEP 2</td><td>Write down the starting time of the experiment. Measure out exactly one liter of water into a drinking container. The idea is to drink this liter of water within 20 minutes.</td></tr>
<tr><td>STEP 3</td><td>Exactly 20 minutes after you started drinking the water, and at intervals of exactly 20 minutes over the next two hours, void your bladder and *measure the volume of urine produced*. Then, *measure the specific gravity* of each sample. Be sure to rinse out the measuring beaker after each sample. Continue the experiment longer than two hours if you are still producing a lot of urine. Record your results in the Data Table.</td></tr>
<tr><td>STEP 4</td><td>When you have finished the experiment, put the urinometers and collecting beakers into the bleach barrels. If you are doing the experiment at home, please wash out all equipment with a mild bleach solution before returning it to school.</td></tr>
</table>

DATA TABLE

SAMPLE	SPECIFIC GRAVITY	URINE VOLUME	URINE PRODUCTION RATE (Divide Volume by 20)
Starting Sample		x x x	x x x
20 min.			
40 min.			
60 min.			
80 min.			
100 min.			
120 min.			

Total Urine Volume = _____ ml

Divide the total urine volume by the amount of water consumed to get the percent of water returned:

$$\frac{\text{Total Urine Volume}}{\text{Water Consumed}} \times 100 = \textit{Percent Returned} = \text{_____} \%$$

ACTIVITY #3

"OPTIONAL WATER BALANCE STUDY"

Your body maintains its *water balance* by ensuring that the water intake equals the water loss during the day. In order to do a complete analysis of water balance we would have to measure fluids lost by skin evaporation, breathing, and in the urine and feces. The fluid intake includes both the water you drink and the fluid content of the foods you eat.

A complete water balance study is beyond the practical limits of this course, but useful information can be obtained by measuring only fluids consumed and the urine produced. During this investigation you will measure the volume of all fluids consumed (milk, soft drinks, coffee, water, etc.) and the volume of urine every time you void your bladder during a 24-hour day.

GO GET

1. A urine collection measuring beaker.

2. pH test paper kit (if available).

NOW

1. Record the volume and time of day for all the fluids you consume during the 24 hour period. Conventional measuring units can be easily converted into ml for the data table.

>   1 ounce = 30 ml
>   1 cup = 237 ml
>   1 pint = 473 ml
>   1 quart = 946 ml

2. Your instructor may provide you with pH test paper. The pH of urine is sometimes affected by the kinds of foods you eat, and this aspect of urine chemistry can easily be tested by dipping a strip of pH test paper into each sample, then reading the color code for the pH.

3. Record your starting time. It's best to start at a familiar time, like when you get up in the morning. Void your bladder, then begin. *Remember:* The urine you just voided was produced the night before.

4. Keep track of every volume of liquid you consume over the next 24 hours. Measure the amount of urine you produce (and its pH) until the starting time on the next day.

5. Have fun! It's quite an adventure being both the experimental animal and the scientist.

DATA TABLE

Time of Day	Description of Fluid Consumed	Amount of Fluid Consumed	Amount of Urine Produced	pH of Urine
WATER BALANCE STUDY				

Total Fluid Consumed in 24 Hours = _____ ml

Total Urine Produced in 24 Hours = _____ ml

$$\frac{\text{Urine Produced}}{\text{Fluid Consumed}} \times 100 = \textit{Percent Returned} = \text{_____} \%$$

Summary Questions

1. What are the three functions of the kidney?

2. Where in the kidney is the blood filtered?

3. What is different about the pressure of blood entering the kidney?

4. The ureter is a _____ that connects the _____ to the _____ .

5. What is the name of the tube between the bladder and the outside?

6. During the water-loading experiment, what was the relationship between the rate of urine produced compared to its specific gravity?

7. Why would high blood pressure make a person a possible condidate for a kidney transplant?

8. Describe the relationship between urine production and dehydration.

KIDNEY FUNCTION REVISITED

How many times, and in how many places, has it been nearly impossible to find a bathroom when you really have to pee? America can't make a theft-proof car, but a locked bathroom is impenetrable even to the U.S. Marine Corps. And where can you get a drink of fresh water when you're really thirsty? Ever wonder why there are so many kidney transplants in a country that talks only about heart problems? Our society mostly ignores the importance of drinking plenty of fresh water and the unhealthy habit of retaining urine in the bladder all day.

During last week's lab you drank a lot of water and urinated for two hours. This week you will analyze the data you collected and consider some of the possible implications. But before doing that, let's begin with a general survey of the urinary system.

ACTIVITIES

ACTIVITY #1

"THE URINARY SYSTEM"

Your urinary system is designed to slowly clean and balance the fluid part of your blood. During this processing, the waste products from metabolism are removed, and the electrolytes (salts), fluids, acids, and bases are adjusted in their concentrations so that they are in the proper balance for the cells of the body. The urinary structures include: *kidneys*, *ureters*, *bladder*, and *urethra*.

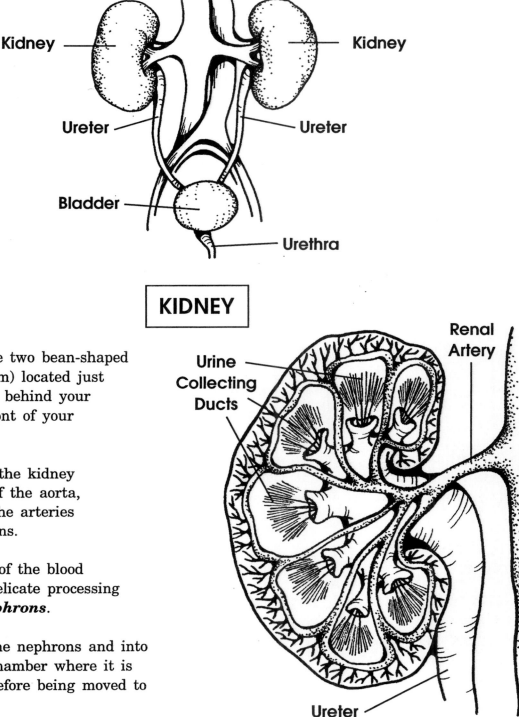

KIDNEY

Your kidneys are the two bean-shaped organs (10 cm x 5 cm) located just above your waistline behind your gut cavity and in front of your thick back muscles.

The artery entering the kidney is a direct branch off the aorta, and is bigger than the arteries supplying other organs.

Fluid is filtered out of the blood and flows through delicate processing structures called *nephrons*.

Urine flows out of the nephrons and into a kidney collection-chamber where it is temporarily stored before being moved to the bladder.

174

There are a million or so nephrons packed together in the outer half of each kidney.

As fluid passes through the nephron tubules, it is processed in a very special way. Cell transport mechanisms ensure that the waste products stay in the tubules, while the needed salts, acids and bases, water, and nutrients are returned to the blood supply.

The small amount of remaining fluid with waste products is collected and passed on as *urine* to the bladder.

NEPHRON

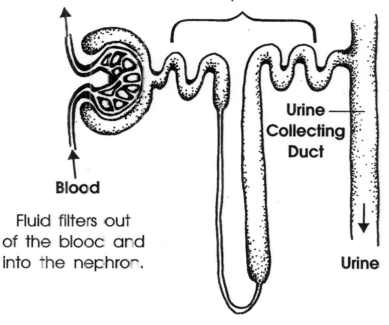

Fluid is processed.

Blood

Fluid filters out of the blood and into the nephron.

Urine Collecting Duct

Urine

? QUESTION

1. Because of the potential for infection, it is a surgical risk to enter into the cavity that actually contains the intestines. If the doctor does not want to cut through this cavity to get to the kidney, then what is the other surgical option?

2. In the Conditioning Lab you investigated the influence of vessel diameter on blood flow. What blood pressure (higher or lower) would you expect in the kidney as compared to other organs in the body? Explain.

3. The blood pressure in the arteries supplying most body organs is low enough to ensure that fluid is not pushed out of the blood. Based on your answer to question #2, what would you expect to happen as blood passes through the kidney?

4. Typically, you will urinate more when you've had a very upsetting day. What would cause more urine to be produced on those days?

5. A man who had high blood pressure for several years was talking with his doctor. He asked the doctor if his high blood pressure would make him a candidate for a heart transplant. The doctor quickly replied, "No, a kidney transplant." Explain this answer. *Check your answer with your instructor.*

A sectioned kidney from the demonstration table.

Kidney

NOW

Examine the kidney, and see if you can identify the *processing area* where the nephrons are located and the *collecting area* where the urine accumulates before being sent to the bladder.

Make a quick sketch of the kidney.

URETERS

Each kidney has a tube connecting it with the bladder. This tube, called the **ureter**, is about 25 cm long and 1–2 cm in diameter.

The walls of the ureter are capable of a special muscular contraction that "milks" the urine to the bladder. This type of contraction, called *peristalsis,* is similar to the process that moves food through your digestive tract.

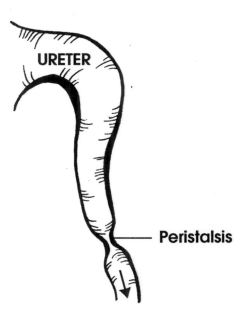

URETER

Peristalsis

? QUESTION

1. What do you think would happen to the muscle contractions of the ureters if they were repeatedly damaged by frequent bladder infections?

2. What surgical procedure would have to be done if the injured ureters could not adequately move urine from the kidney to the bladder?

BLADDER

The bladder is about the size of your fist, and is located just behind the pubic bone and below the gut cavity. In women, the bladder is just under the uterus and in front of the vagina; in men, it is in front of the rectum (lower part of the bowel). The **bladder** is a muscular storage chamber that can hold $\frac{3}{4}$ of a liter of urine, although you will have a strong urge to urinate when it is filled with only half that amount.

? QUESTION

1. Usually patients are back to work much sooner after bladder surgery than after kidney surgery. Assume that both the bladder and the kidney require the same amount of time to heal. Why is there such a difference in time before returning to work? **Hint:** See Question #1, under the kidney discussion.

2. A woman who is four months pregnant has been getting up twice during the night to urinate. What can you tell her about the anatomical cause of her problem?

URETHRA

The tube from the bladder to the outside of the body is called the **urethra**. In women, it is a short direct tube exiting in front of the vagina. In men, its route is not as direct—it passes through the prostate gland at the base of the bladder, and then through the penis to the outside of the body.

? QUESTION

1. In which sex is it easier for intestinal bacteria to cause an infection in the urethra? Explain.

2. Doctors lobbied to get a law passed that required all public bathrooms to have a toilet seat that is open in the front (shaped like a horseshoe). For whom did they do this? (circle your choice.)

    Adult Males   or   Young Boys   or   Adult Females   or   Young Females

    Explain your answer.

3. It is typical for men to develop some enlargement of the prostate gland after 40 years of age. A routine question doctors ask their male patients is: "Do you have any trouble starting to urinate, or is the flow less than before?" Explain their reasons for asking this question.

ACTIVITY #2

"ANALYSIS OF WATER LOADING EXPERIMENT"

Refer to the experimental results recorded in the Data Table of Activity #2 in last week's kidney lab.

URINE PRODUCTION AND SPECIFIC GRAVITY

The data for urine production rate and urine specific gravity can be plotted on the "Water Loading Experiment" graph on the next page.

| STEP 1 | The left vertical axis of the graph has the values for urine production rates. Use a circle (o) to mark each of the six 20-minute samples. |

| STEP 2 | The right vertical axis of the graph has the specific gravity values. Use an "x" to mark each of the seven samples. (Include the urine sample before you drank the one liter of water.) |

| STEP 3 | Draw a "best-fit line" through the urine production rates. |

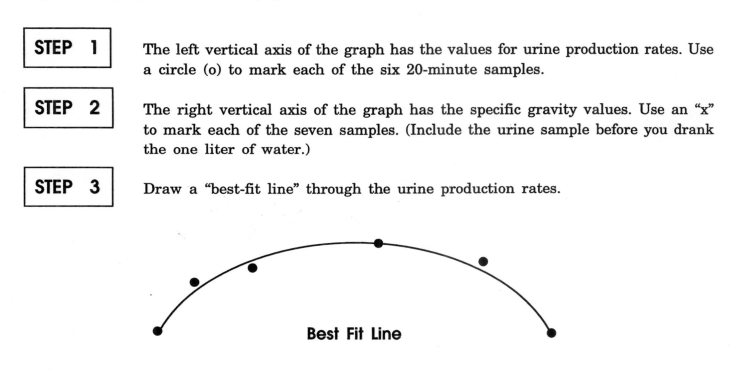

Best Fit Line

| STEP 4 | Draw a "best-fit line" through the specific gravity values. |

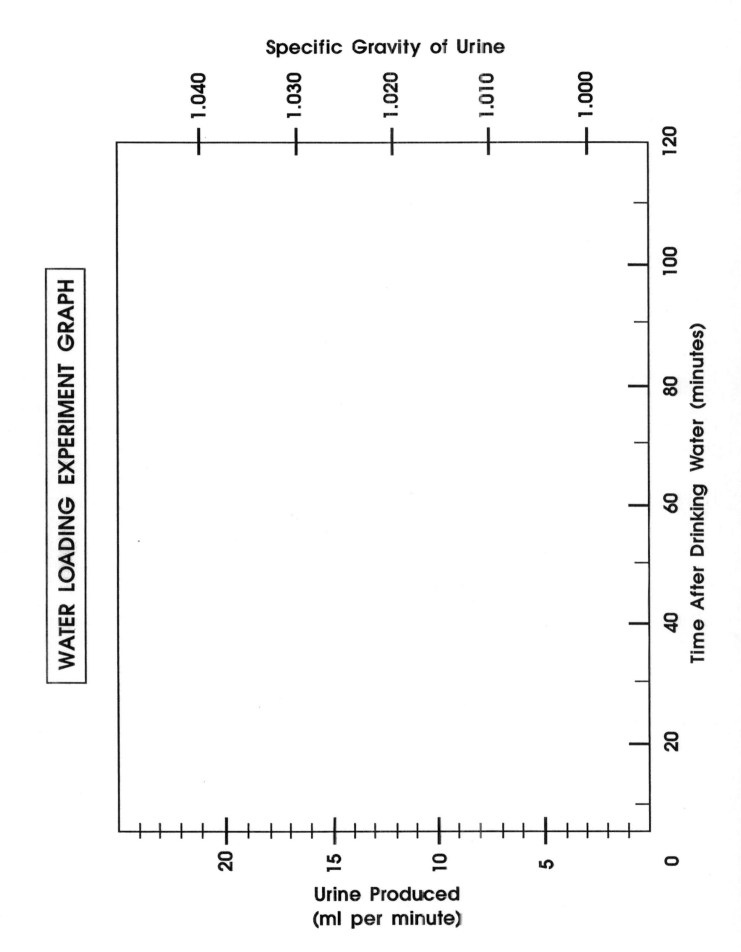

WATER LOADING EXPERIMENT GRAPH

Specific Gravity of Urine

1.040   1.030   1.020   1.010   1.000

Time After Drinking Water (minutes)

0   20   40   60   80   100   120

Urine Produced
(ml per minute)

20   15   10   5

1. What happened to your urine production rate during the first hour after you drank the liter of water?

2. There is a very dense network of blood vessels in the wall of the small intestine. What is the purpose of those vessels?

3. When you drank the liter of water it went quickly from the stomach into the intestine. Where is the very next place that water goes?

4. What happens to your blood pressure during the first hour after you drink a liter of water? *Hint:* Refer to Question #3.

5. During this two-hour experiment, you produced a lot of urine. What happens to your blood pressure after the urine is produced?

6. What happened to the specific gravity as your urine production rate increased?

7. What is the kidney getting rid of?

TIME OF PEAK URINE PRODUCTION

One important function of your kidneys is revealed by measuring the amount of time required to reach maximum output of urine after water-overload to the system. This is testing the ability of the kidney to return the body's water balance to normal. Some kidneys are better than others in this particular function.

1. Refer to the water-loading graph. Determine how many minutes it took until your kidneys reached the peak (maximum) of urine production rate.

_____ minutes required to reach peak urine production

2. Determine the average peak production time for everyone in the lab class.

_____ average peak production time for your lab class

? QUESTION

1. How does your peak production time compare to the class average?

2. If you have a faster peak production time, do you notice that you have to pee sooner than most people when you drink coffee or beer?

Explain.

3. If you have a slower peak production time, do you notice that you have to get up at night to pee when you drink coffee or beer earlier in the evening?

Explain.

PERCENT RETURNED

The **percent returned** $\left(\frac{\text{urine volume}}{\text{water consumed}}\right)$ at the end of the water-loading test is an indication of whether or not the person was dehydrated before the experiment began. Minor dehydration is not necessarily a bad thing, but it does suggest that your kidneys must do more work to maintain the water balance. Most people in our society don't drink enough water and are partially dehydrated.

NOW

Compare your *percent returned* to others in your lab class.

_____ = highest % returned in the class

_____ = lowest % returned in the class

_____ = my % returned

1. Based on the experiments you have analyzed during this lab, do you think that you will start drinking more water during the day?

2. We know that the bladder muscle wall thickens abnormally when a person has a lifetime habit of retaining urine during the day. This physical change typically leads to a condition in which . . . (circle your choice)

the bladder holds
more urine
and eliminates
it completely.

or

the bladder does not hold
as much urine
and does not eliminate
it completely.

Knowing these facts, are you more willing to go to the bathroom when you feel the urge?

3. If you don't follow through on your decision to pee more often, what condition might you develop later in your life?

ACTIVITY #3

"ANALYSIS OF WATER BALANCE STUDY"

If you did the optional 24-hour water-balance study, then refer to your data.

NOW

1. On the back of this page make a bar chart (like this one) of your results.

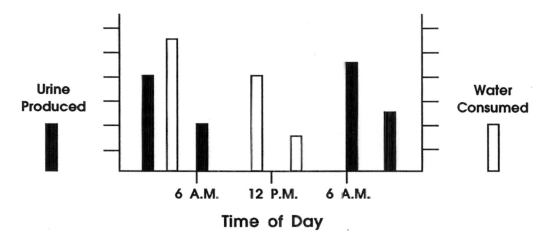

2. Calculate the *percent returned* for the 24 hour period:

$$\frac{\text{Total Urine Volume}}{\text{Total Fluids Consumed}} \quad x \quad 100$$

3. In a few paragraphs, summarize your observations and conclusions about this water-balance study.

SAMENESS & VARIETY

Summary Questions

1. Define mitosis and meiosis.

2. Explain why living organisms need both.

3. In all cases, is it you or your DNA that reproduces? Explain.

4. Define homologous pairs.

5. Write a one sentence description of each phase of mitosis: prophase, metaphase, anaphase, and telophase.

6. Define synapsis.

7. Define crossing-over and explain how "new mixes" are created from parent chromosomes.

8. Discuss how independent assortment and random fusion create genetic variety in offspring.

SAMENESS AND VARIETY

Is it better to be the same as everyone else?
or
Is it better to be different?

We struggle with these questions in our personal lives. Would it surprise you to learn that all life, in terms of its reproductive strategy, has struggled with the same basic questions? It might also surprise you to learn that there is no *one* answer to reproduction, but *two* answers.

The strategy of producing offspring that are genetically the same as the parent is called **asexual reproduction,** and is accomplished through a cell division process termed **mitosis**. Asexual reproduction is the simplest and oldest form of reproduction, and it relies on a single parent.

The strategy of producing offspring that express genetic variety is termed **sexual reproduction**. It is accomplished through a cell division process called **meiosis** and a fusion process called **fertilization**. Sexual reproduction is complex and it usually relies on two parents.

Many organisms have lost the means to reproduce asexually except for cell replacement or growth. But some less specialized species use both modes, depending on the time of year. The one certainty is that organisms exist today only because they have incorporated both sameness and variety in their struggle to live and reproduce.

ACTIVITIES

ACTIVITY #1

"DNA REPLICATION"

During the last 40 years, the study of biochemistry revealed a fact that stunned the scientific establishment and transformed our approach to biology.

The fact is this: ***A chemical called DNA reproduces—not the individual.***

Also, it was discovered that the sorting of DNA by the cell is what controls the *sameness* or *variety* of the next generation. This does not mean that the organism isn't important. The organism houses the cells that contain the DNA molecules. Without the organism, DNA would destabilize and fall apart. But the unit that actually gets passed on to the next generation is the DNA. That's why the discussion of reproduction is centered around this unique molecule.

So, first let's take a look at the structure of DNA and how it reproduces itself. Then, we will see how DNA uses an organism to achieve sameness by asexual reproduction, or achieves variety by sexual reproduction.

DNA STRUCTURE

DNA is an extremely thin, long, ladder-like molecule that has two "rails" made of sugar and phosphate, and many "rungs" made of special complementary bases.

The DNA bases (rungs) are molecular units that combine only in two kinds of pairs.

The base ***adenine*** (A) is always paired with ***thymine*** (T), and ***cytosine*** (C) is always paired with ***guanine*** (G). Therefore, if you know *one* of the complementary bases, you can easily figure out the other.

The term ***nucleotide*** is used as a name for the repeating subunits in a DNA molecule. A nucleotide is actually a phosphate, a sugar, and a base hooked together as a basic building unit.

DNA Molecule

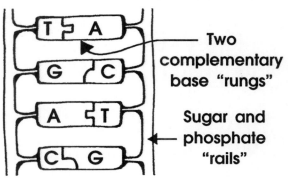

Two complementary base "rungs"

Sugar and phosphate "rails"

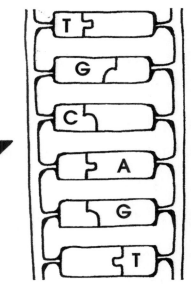

? QUESTION

1. Guanine always pairs with _____.

2. Thymine always pairs with _____.

3. Fill in the complementary nucleotides on this DNA ladder.

REPRODUCTION OF DNA

Before DNA can copy itself the cell must make lots of extra A, T, G, and C. Then the DNA unzips between the two bases and adds nucleotides to each side of the unzipped DNA molecule.

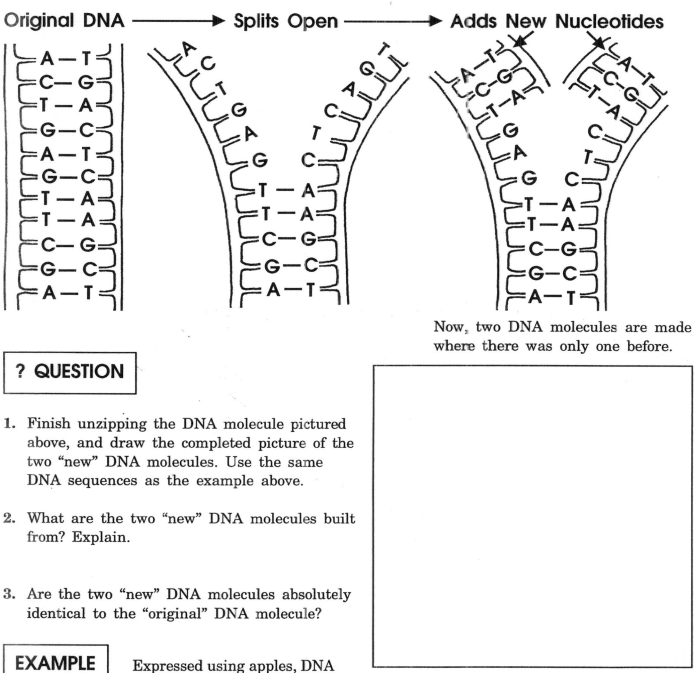

Original DNA ——→ Splits Open ——→ Adds New Nucleotides

Now, two DNA molecules are made where there was only one before.

? QUESTION

1. Finish unzipping the DNA molecule pictured above, and draw the completed picture of the two "new" DNA molecules. Use the same DNA sequences as the example above.

2. What are the two "new" DNA molecules built from? Explain.

3. Are the two "new" DNA molecules absolutely identical to the "original" DNA molecule?

EXAMPLE Expressed using apples, DNA replication looks like this:

Look at the model of DNA on the demonstration table.

We have been presenting DNA as a straight ladder, but actually it is twisted on itself like a spiral staircase. This shape is called a *helix*.

THE CHROMOSOME

Normally DNA exists as loose strands (*chromatin*) in the nucleus of a cell. This nuclear DNA sends a message (RNA) to the ribosomes where protein and enzymes are synthesized.

When stretched out, the length of one DNA molecule in a human cell is almost 4 cm. However, the cell itself is but a tiny fraction of that size.

Problem: During cell reproduction the DNA must be able to move around. So it shortens its length by tightly coiling up. In doing so, the DNA strands become wider and are visible under a microscope. Visible DNA is called a *chromosome*.

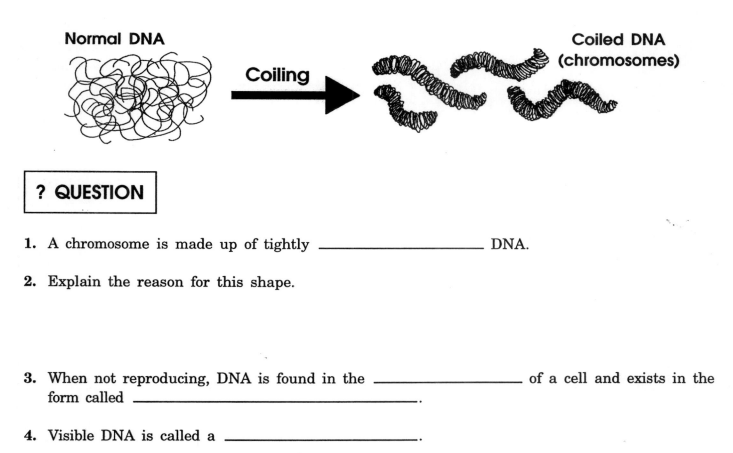

Normal DNA   Coiling   **Coiled DNA (chromosomes)**

? QUESTION

1. A chromosome is made up of tightly _____ DNA.

2. Explain the reason for this shape.

3. When not reproducing, DNA is found in the _____ of a cell and exists in the form called _____ .

4. Visible DNA is called a _____ .

ACTIVITY #2

"CHROMOSOME SETS"

The number of chromosomes in a cell varies from species to species, but it is exactly the same among individual members of the same species.

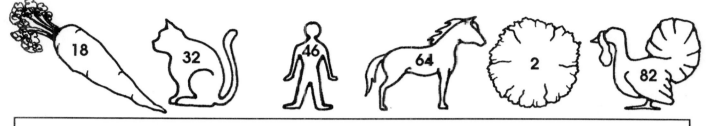

THE ONE-SET CONCEPT OF CHROMOSOMES: WHAT IS HAPLOID?

All species have one or more sets of chromosomes. This means that chromosomes come in *sets*, and the *number* of chromosomes in a set depends on the particular species.

A set of chromosomes includes *one copy* of all of the genes necessary to control the biochemical activities of a species. Most species have either one or two chromosome sets.

In genetics a single set of chromosomes is symbolized by the letter "**n**." Any cell that has only one set of chromosomes is termed **haploid**. Haploid means that the cell has *one* of each *kind* of chromosome.

? QUESTION

The set concept will be used throughout the rest of this lab, so the following questions were designed to aid you in understanding and recognizing sets. Remember that a set is a group of objects related in function and generally used together.

1. Pretend that the *fingers* of one hand represent chromosomes. (Count your thumb as a finger.) Hold up your hand.

   a. How many fingers (chromosomes) do you have on one hand? _____

   b. Do you have different kinds of fingers on one hand? _____

   c. Do you have more than one of each kind of finger on one hand? _____

   d. Judging by the definition of a set, you have _____ set(s) of fingers on one hand.

2. Pretend that all the numbers within the circle represent chromosomes.

   a. How many numbers are there? _____

   b. Are there different kinds of numbers? _____

   c. Is there more than one of each kind of number? _____

   d. Judging by the definition, you have _____ set(s) of numbers.

3. Pretend these lines represent chromosomes.

    a. How many lines are there? _____

    b. Are there different kinds of lines? _____

    c. Is there more than one of each kind of line? _____

    d. Judging by the set definition, you have _____ set(s) of lines.

       Answers:

#1	#2	#3
a=5	a=7	a=4
b=yes	b=yes	b=yes
c=no	c=no	c=no
d=1	d=1	d=1

4. Explain the one-set concept of chromosomes.

THE TWO-SET CONCEPT OF CHROMOSOMES: WHAT IS DIPLOID?

Simple organisms and the gametes of complex organisms are haploid. That is, they have a single set of chromosomes. Complex organisms require *two sets* of chromosomes to survive. We will discuss the details of this two-set requirement in a later lab.

In genetics, the two-set condition is symbolized as "**2n**," and is called *diploid*. Diploid means that the cell has *two* of each *kind* of chromosome.

? QUESTION

1. Pretend that the fingers of *both* of your hands represent chromosomes. Hold up your hands.

    a. How many fingers (chromosomes) do you have?

    _____

    b. Regarding both hands, do you have a *duplication* of each of the *kinds* of fingers? _____

    c. How many sets of fingers do you have?

    _____

    d. How many fingers are in a single set?

    _____

    e. Draw a simple sketch of a single set of fingers.

2. Pretend that all the numbers within the circle represent chromosomes.

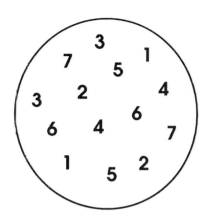

    a. How many total numbers are there? _____

    b. Is there a duplication of each kind of number? _____

    c. How many sets of numbers are there? _____

    d. How many numbers are in each set? _____

    e. Draw a sketch of a single set of numbers.

3. Pretend that the lines within the circle represent chromosomes.

    a. How many lines are there? _____

    b. Is there a duplication of each kind of line? _____

    c. How many sets of lines are there? _____

    d. How many lines are there in each set? _____

    e. Draw a sketch of a single set of lines.

Answers:

#1	#2	#3
a=10	a=14	a=8
b=yes	b=yes	b=yes
c=2	c=2	c=2
d=5	d=7	d=4

4. Explain the two-set concept of chromosomes.

193

HOMOLOGOUS CHROMOSOMES

The word *homologous* means "the same," and the term **homologous chromosomes** refers to the pairs of chromosomes in a diploid cell that carry genes for the same traits. We will discuss more details about cell divison in the following Activities, but for now, keep this note in mind: *Each chromosome of a homologous pair comes from a different parent.* Humans are diploid and have 46 chromosomes (two sets of 23). This means that we have 23 homologous chromosome pairs.

Although both chromosomes of a particular homologous pair carry the *same genes*, these genes may be slightly different *forms*. For example, one might be the "blue eye" form, and the other might be the "brown eye" form. (More about that later.)

? QUESTION

1. Pretend that the fingers of both of your hands are chromosomes. Hold up both your hands.

    a. How many individual fingers (chromosomes) do you have? _____

    b. How many homologous pairs are there? _____

    c. How many sets of fingers do you have? _____

    d. How many homologous pairs are in one set of fingers? _____

2. Pretend that the numbers in the circle are chromosomes.

    a. How many individual numbers are there? _____

    b. How many homologous pairs of numbers are there?

    _____

    c. How many sets of numbers are there? _____

    d. How many homologous pairs are in one set of numbers? _____

3. Pretend that all the lines within the circle are chromosomes.

    a. How many individual lines are there? _____

    b. How many homologous pairs of lines are there? _____

    c. How many sets of lines are there? _____

    d. How many homologous pairs of lines are in one set? _____

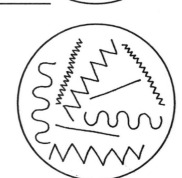

Answers:

#1	#2	#3
a=10	a=14	a=8
b=5	b=7	b=4
c=2	c=2	c=2
d=0	d=0	d=0

4. Explain in simple terms what homologous chromosomes are.

ACTIVITY #3

"ASEXUAL REPRODUCTION OF CELLS—MITOSIS"

Asexual reproduction of cells is called *mitosis*. Immediately before this cell division process begins, the DNA of a cell (either haploid or diploid) duplicates itself creating two identical copies of every DNA molecule (and chromosome). The DNA copies move to opposite ends of the cell. Then the cell partitions itself into two cells (each with *exactly* the same DNA as the original cell). The purpose of asexual reproduction by mitosis is to create new cells that are genetically *identical* to the original cell.

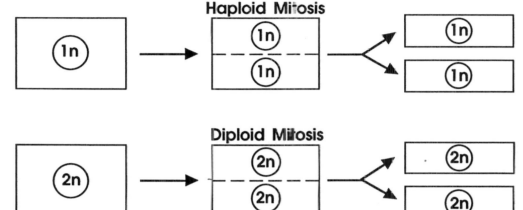

PROBLEM

1. Let's imagine what would happen to the amount of DNA material in a cell if, when it reproduced, it was *halved* instead of duplicated. Complete this cell box.

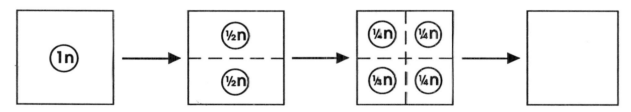

2. Explain what this mistake creates in the new cells.

FINALLY

As stated above, some cells start with *one set* of DNA molecules (**1n**), and other cells start with *two sets* of DNA molecules (**2n**). For a dividing cell to maintain its original set # in the new cells, it must duplicate its genetic material prior to beginning mitosis. In addition, the genetic material must be divided in such a way that no new cell is missing any DNA, or has more DNA than the original cell.

Whatever amount of DNA the original cell has prior to mitosis, its offspring cells must end with after the process is complete.

1. When the process of mitosis is used for organism reproduction, are the new organisms exact genetic duplicates of the parent organism?

2. If organisms use mitosis for reproduction, would their offspring exhibit *sameness* or *variety*?

3. If asexual reproduction produces identical offspring, then how does such an organism "change" over time?

4. What would be the *advantage* of reproducing asexually?

5. What would be the *disadvantage* of reproducing asexually?

6. Starting with the *haploid* cell below, draw the next *three* generations of that cell as it reproduces.

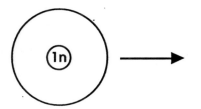

7. Starting with the *diploid* cell below, draw the next *three* generations of that cell as it reproduces.

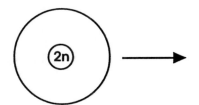

ACTIVITY #4

"CHROMOSOME MOVEMENT DURING MITOSIS"

Chromosomes move during mitosis. They replicate themselves (see Activity #1), and the copies separate, allowing two cells to be created from one. This movement of DNA material in the form of chromosomes (coiled DNA) has several "phases," which are described in this Activity.

GO GET

1. One red and one yellow crayon.

2. A package of chromosome beads. Each package should contain 8 chromosomes.

NOW

The human has 46 chromosomes (23 homologous pairs). We will follow the movements of 4 chromosomes (2 homologous pairs) as an example of what all the chromosomes are doing during mitosis.

Start with 4 of the chromosomes from your package. This is how a cell would look *prior* to duplicating its genetic material and undergoing the process of mitosis:

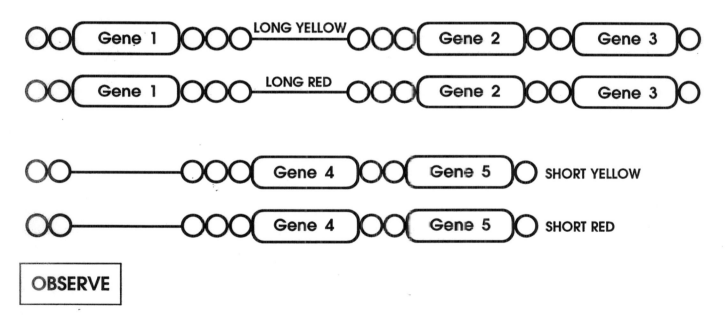

OBSERVE

1. The two short chromosomes represent *one* homologous pair, and the two long chromosomes represent *another* homologous pair. The *red* set represents the chromosomes from your mother, and the *yellow* set is the chromosomes from your father.

2. Color the chromosome beads above with your crayons, and as you go through this Activity, use the crayons to help you keep track of the chromosomes that came from your father and from your mother.

3. The bead chromsomes are labeled A^1, A^2, B^1, and B^2 in the next section.

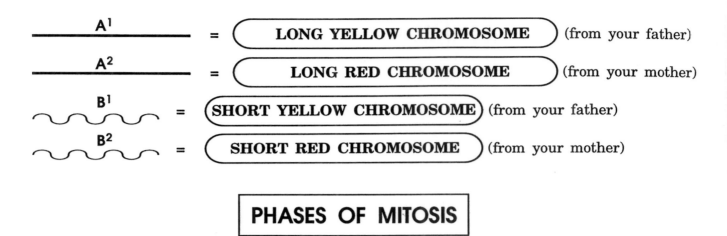

A^1 = LONG YELLOW CHROMOSOME (from your father)

A^2 = LONG RED CHROMOSOME (from your mother)

B^1 = SHORT YELLOW CHROMOSOME (from your father)

B^2 = SHORT RED CHROMOSOME (from your mother)

PHASES OF MITOSIS

Biologists sometimes describe mitosis as having several phases. Ask your instructor if you are required to memorize the names of the phases. If so, remember the phrase: **Pay Me Any Time.** This will help you to remember the sequence of phases in mitosis.

Pay attention to the different events as they occur in each phase, and mimic the phases by using your bead chromosomes. Interphase is usually considered to be the stage *before* mitosis actually begins. We include it as part of the mitosis discussion, but your textbook will say that mitosis begins with prophase.

1. **Early Interphase:** Although we have diagramed the DNA as long, thin chromosomes, in reality it is not coiled up yet, and is *not* visible as chromosomes until **prophase**. However, it is best to label the DNA at this stage so that we can remember what the parent cell starts with.

2. **Later Interphase:** Each DNA molecule has duplicated itself.

   We have diagramed these "doubled" DNA molecules as though we could see them. Actually, DNA is still in the long, thread-like form.

   Duplicate your beads now, using the other four bead chromosomes from the package.

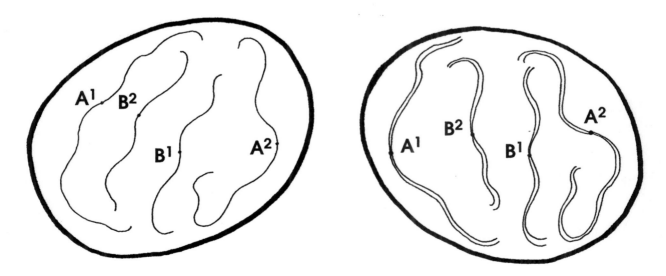

3. **Prophase:** This is when the DNA coils up and the chromosomes are now *visible* under the microscope.

   Each chromosome is now doubled, and consists of two absolutely *identical* "chromatids." (A **chromatid** is the name for one of the duplicated DNA molecules that has coiled itself into a chromosome form and is attached to the other chromatid.)

4. **Metaphase:** The chromosomes (each consisting of two chromatids) line up end to end, in any order, along the *midline* of the cell. Spindle fibers have formed and are attached to the chromosomes. You will have to imagine fine threads attached to your beads.

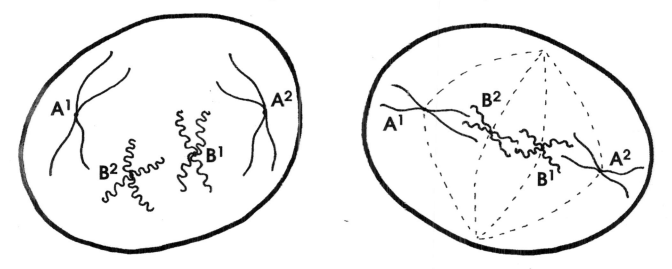

5. **Anaphase:** The spindle fibers pull the duplicated chromatids apart and move them to opposite ends (poles) of the cell.

6. **Telophase:** Chromosomes are at opposite ends of the cell, and the cell divides into two cells.

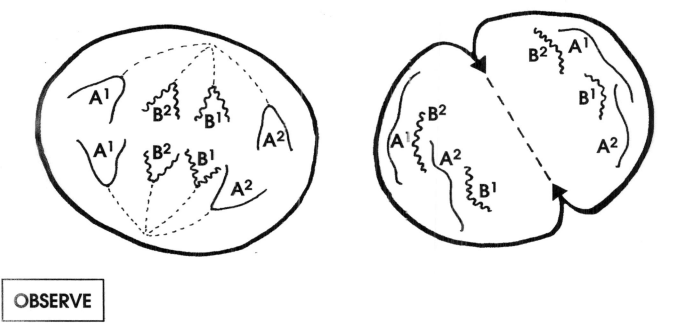

| OBSERVE |

Notice that you started with a cell having two sets of chromosomes, and you ended with two cells, each having two sets of chromosomes.

199

1. Are the two groups of chromosomes at *telophase* identical to the group of chromosomes you started with in *interphase* prior to DNA duplication?

2. Are the new cells identical to the original cells?

3. What is the name for this cell division process?

4. What kind of reproduction is it?

5. Is mitosis going on in your body right now?

   What kind of cells are you producing by this process?

6. Name two situations where your body must reproduce cells by *mitosis*.

   a.

   b.

7. What do you think might be going on as you age (get wrinkles, grey hair, lose your hair, etc.)?

ACTIVITY #5

"ONION ROOT TIP"

It's time to review what you learned in Activity #4.

You will use a microscope to find the various phases of mitosis (cell division) in the root tip of an onion. Put on your investigator's hat and search for the "clues" that reveal each phase.

GO GET

1. A compound microscope.

2. A prepared slide of an onion root tip.

NOW

1. Under low power, note that the root tip is covered by a root cap (like a thimble over your finger). Behind the root cap is an area of square-shaped cells that are undergoing cell division.

2. Look at this area under the high power (430x). If the cells are *rectangular,* then you are in the wrong place.

3. *Find every stage of mitosis.*

Onion Root Tip

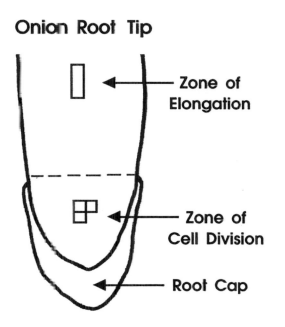

Zone of Elongation

Zone of Cell Division

Root Cap

201

4. Draw a simple sketch of what you see at each phase of mitosis.

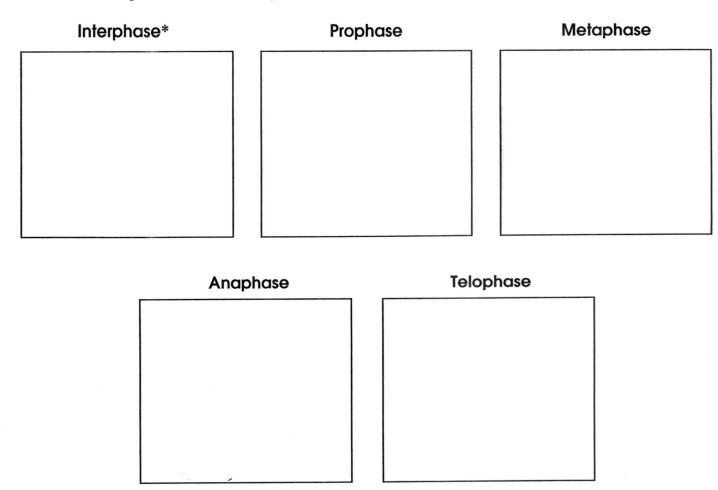

| Interphase* | Prophase | Metaphase |

| Anaphase | Telophase |

* Interphase will look like a stained nucleus. You won't be able to see the DNA threads.

5. Optional: (Ask your instructor if you are to do this experiment.) There is a way that you can estimate the relative amount of time that a cell spends in each phase of the cell cycle. Count all of the cells in the zone of cell division and record how many of them are in each phase. Then figure what percent each stage is of the total. This is an indication of the relative time a cell spends in each stage of cell division. Does this make sense to you? Do it, it will.

Phase   **# of Cells in Phase**   **% of Total Cells**

ACTIVITY #6

"SEXUAL REPRODUCTION"

Sexual reproduction is the process of creating *variety* in the offspring of a species. It consists of two parts: **meiosis** and **fertilization**.

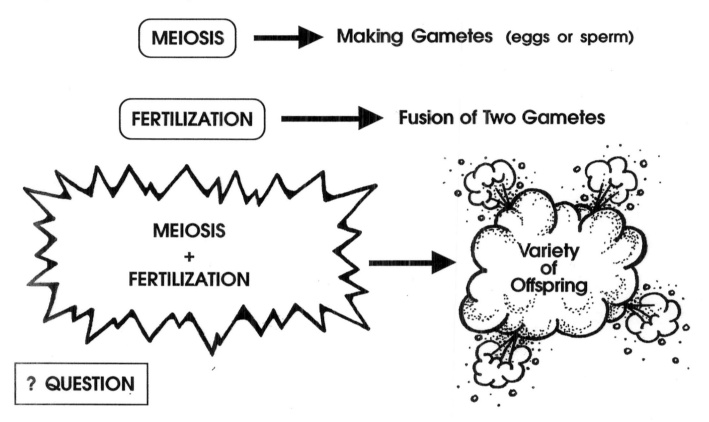

MEIOSIS ⟶ Making Gametes (eggs or sperm)

FERTILIZATION ⟶ Fusion of Two Gametes

MEIOSIS + FERTILIZATION ⟶ Variety of Offspring

? QUESTION

1. What is the advantage of producing variety in offspring?

2. What is the disadvantage of producing variety in offspring?

SET CHANGES DURING MEIOSIS AND FERTILIZATION

Review Activity #2 if you have forgotten what a "set" is.

MEIOSIS

Meiosis starts with a single *diploid* cell and ends with four *haploid* cells.

The original cell has two sets (2n) of DNA molecules which are then duplicated. After DNA duplication, the original cell divides *twice,* producing *four* cells—each with a *single* set (1n) of DNA molecules (chromosomes).

These four cells are haploid and their chromosomes have been mixed in such a way as to produce *genetic variety*. We will present some of the details of that process in Activity #7.

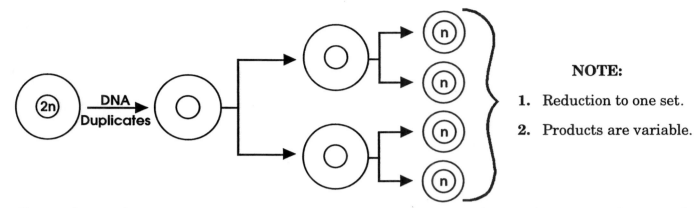

NOTE:

1. Reduction to one set.

2. Products are variable.

The product cells of meiosis are called **gametes**, and these can fuse with gametes from another organism of the same species to begin the next generation.

FERTILIZATION

Fertilization is the fusion of *two haploid gametes*. It results in the chromosome set number returning to 2n. This allows the next generation of the species to have the same "set" number of chromosomes as the parent generation.

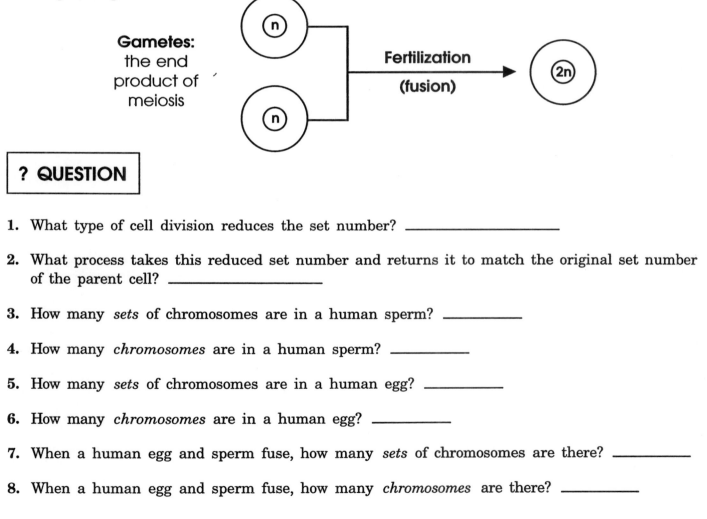

? QUESTION

1. What type of cell division reduces the set number? _____

2. What process takes this reduced set number and returns it to match the original set number of the parent cell? _____

3. How many *sets* of chromosomes are in a human sperm? _____

4. How many *chromosomes* are in a human sperm? _____

5. How many *sets* of chromosomes are in a human egg? _____

6. How many *chromosomes* are in a human egg? _____

7. When a human egg and sperm fuse, how many *sets* of chromosomes are there? _____

8. When a human egg and sperm fuse, how many *chromosomes* are there? _____

ACTIVITY #7

"THE EVENTS THAT CREATE VARIETY"

The essence of meiosis is the production of haploid cells from diploid cells. The essence of fertilization is the recombination of two haploid gametes to produce the next diploid generation. During meiosis and fertilization there are *three* events that create *genetic variety* in the next generation: **crossing-over, independent assortment,** and **random fusion of gametes.**

None of these genetic variety events would be possible without a very special process during meiosis called **synapsis**. This is the single most critical event that makes meiosis so different from mitosis.

SYNAPSIS

Synapsis is defined as the "pairing up" process of *homologous pairs* early in meiosis.

Let's illustrate the differences between mitosis and meiosis using the same chromosomes from our earlier mitosis diagrams.

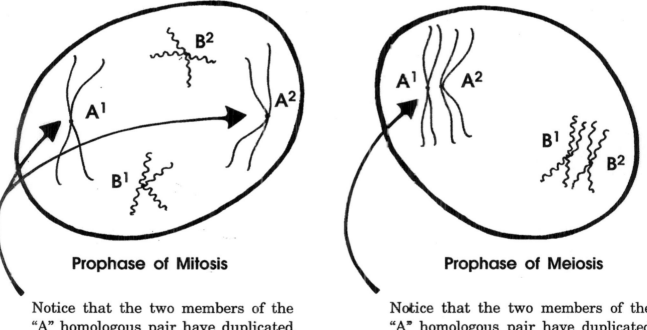

Prophase of Mitosis

Prophase of Meiosis

Notice that the two members of the "A" homologous pair have duplicated themselves, and they will be moved around the cell *separately* from each other.

Notice that the two members of the "A" homologous pair have duplicated and they have "paired up" *(synapsis)*. This paired grouping is called a **tetrad** (meaning four chromatids) and they will be moved around the cell *together*.

205

1. The two "A" chromosomes are concerned with the same traits and are called _____ pairs.

2. Are the chromatids of the "A¹" chromosomes identical or different? _____

3. Are the chromosomes "A¹" and "A²" absolutely identical? _____

4. Are the "A" and "B" chromosomes homologous pairs? _____

5. In meiosis, do the "A" and "B" chromosomes pair up with each other (synapse)? _____

CROSSING-OVER

Crossing-over is the exchange of DNA between the four chromosomes (chromatids) of a *tetrad*.

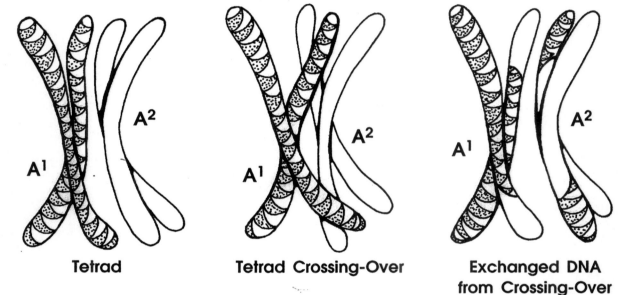

Tetrad   Tetrad Crossing-Over   Exchanged DNA
                                              from Crossing-Over

If you consider that a single chromosome may carry a thousand or more genes, then these small cross-over exchanges are capable of creating hundreds of mixtures of chromosomes. Remember, you received one of the homologous chromosomes of a pair (A²) from your mother and the other (A¹) from your father. The process of crossing-over makes *"new mixes"* of those chromosomes.

? QUESTION

1. What event during meiosis prophase (as opposed to mitosis prophase) makes it possible for crossing-over to occur?

2. As a result of crossing-over, will the "A" chromosome that you pass on to your children be your mother's, your father's, or will it be a mixture of your mother's and father's?

INDEPENDENT ASSORTMENT

During meiosis the tetrads formed by synapsis move around the cell and divide in different ways than we saw in mitosis.

The simplest description of the difference is:
1. the tetrads line up in the middle of the cell (during metaphase), and
2. the cell divides twice, separating the tetrads first into pairs of chromatids and then into single chromosomes, resulting in *four separate cells*.

Each of these cells contains only *one of each kind of chromosome*. They are haploid.

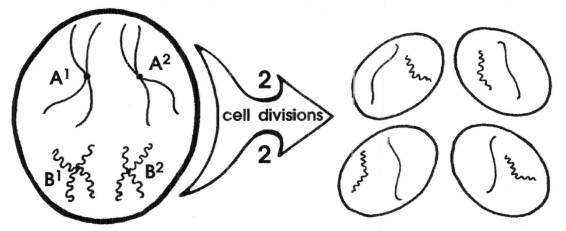

However, there is a very important detail called *independent assortment* that occurs during the chromosome separations of meiosis. Notice that the tetrads are drawn so that the A^1 and B^1 chromosomes are placed on the left side of the cell, and the A^2 and B^2 chromosomes are lined up on the right.

If the tetrads separated equally as drawn here you would see only two kinds of gametes from this process containing the following chromosomes:

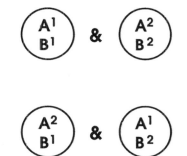

But, if the top tetrad had originally lined up during metaphase with the A^1 chromosomes on the right, then A^2 would have moved with B^1 into a gamete, and A^1 would have moved with B^2 into a gamete as pictured here:

This production of gametes containing different combinations of chromosomes is called ***independent assortment*** because one pair of homologous chromosomes is separated (segregated) into individual gametes independently of how another pair is separated. (A^1 and A^2 have been separated independently of how B^1 and B^2 have been separated.)

IMPORTANT MESSAGE

Remember: Every gamete gets a complete set of chromosomes with only one A chromosome and only one B chromosome.

1. How many *genetically different* gametes were produced in the independent assortment of these two homologous pairs?

2. The human contains 23 pairs of homologous chromosomes, all of which are independently assorted. What do you think the chance would be that one of your gametes would contain either all your mother's or all your father's chromosomes?

RANDOM FUSION OF GAMETES

Two mating individuals have the same kinds and number of chromosomes, but those chromosomes are *not exactly* identical. Because individuals possess different variations of genes, the **random fusion of gametes** (fertilization) from any two individuals will result in more genetic variety in the offspring. Through fertilization the diploid set number is recreated with the offspring receiving one chromosome of a homologous pair from one parent, and the other chromosome of the pair from the other parent.

? QUESTION

To keep the example relatively simple, let's consider only one of the homologous pairs of the human—the A chromosome. (Actually, human chromosomes are referred to by numbers from 1 to 23.)

1. Where did chromosome A^2 come from?

2. Where did chromosome A^4 come from?

A^1 came from
your father

3. What are your possible gametes?

4. What are your mate's possible gametes?

A^3 came from your
mate's father

5. Determine the four possible combinations of your gametes with your mate's gametes. In other words, what genetic variety can we expect in your offspring?

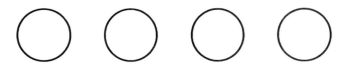

6. Are any of the offspring identical to either parent?

7. What are the three events during *sexual reproduction* that prevent identical children?

209

Summary Questions

1. What are three ways that an understanding of genetics will benefit your life?

2. Define genetics.

3. Define allele.

4. Define homozygous and heterozygous.

5. Define genotype and phenotype.

6. Define dominant and recessive.

7. Define incomplete dominance.

8. Explain how to do a test cross.

9. What are the sex chromosomes?

10. What is a sex-linked trait? Give an example.

11. Does a father give his x-linked genes to his sons, daughters, or both? Explain.

12. Does a mother give her x-linked genes to her sons, daughters, or both? Explain.

GENETICS

INTRODUCTION

More than a hundred years ago Gregor Mendel discovered that hereditary particles are passed from parent to offspring during the reproductive process. These particles were later named **genes**, and the science of studying inheritance was called **genetics**.

Genes can be traced backwards to the very origin of life about 4 billion years ago. When we do so, we find that all traits were new at some point in time, and that a gene's success is determined by natural selection. However, genes do not last forever. And most have already gone extinct.

The investigation into the mechanics of inheritance—the mixing, the passing on, and the function of genes—is one of the greatest scientific puzzles of the 20th century. Understanding genetics has led to the prevention and curing of numerous hereditary diseases. It has substantiated the principle of evolution by natural selection, and has helped human beings to realize their place in Nature's Family Tree.

As a contemporary student, you should note that your individuality is not the result of possessing a trait that no other individual has, but is a result of a particular *combination* of genes. These genes came from your parents and their ancestors before them.

This lab will explore some of the basic principles of genetics, introduce you to basic terminology, and help you apply genetic rules to some hypothetical problems.

ACTIVITIES

ACTIVITY #1

"BASIC TERMINOLOGY"

In order to understand the mechanics of inheritance, we must understand the terminology used to describe this very complex process.

GENES AND CHROMOSOMES

A *gene* is a segment of the DNA molecule and is responsible for manufacturing a protein that either becomes part of the organism's structure or becomes an enzyme that controls biochemical events.

Every organism has a certain number of chromosomes—the exact number depends on the species—and each chromosome is made of many genes. DNA coils up into the form of a chromosome during cell division, and a gene becomes a distinct particle on that chromosome. *Remember:* Chromosomes and DNA molecules are basically the same thing.

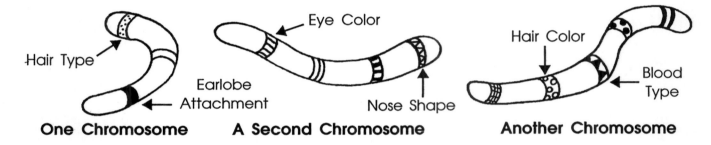

Genes can be described by their exact location on a chromosome. The process of locating genes is called *mapping*. The location of a gene is its *locus,* and geneticists go through great efforts to pinpoint these locations. Knowing where a gene is found on the chromosome is what allows scientists to do genetic research.

? QUESTION

1. The particles that control inherited traits are called _____.

2. These particles are segments of _____, and are responsible for manufacturing a _____ that becomes either _____ or _____.

3. Every living thing on the planet has the same number of chromosomes. _____ (T or F) Explain your answer.

4. Every chromosome has the same identical genes as every other chromosome. _____ (T or F) Explain your answer.

5. The place where you find a particular gene on a chromosome is called the _____.

HOMOLOGOUS CHROMOSOMES

We have discussed homologous chromosomes before. This idea is essential to the understanding of genetics, so we will review it again.

INFORMATION

1. Very simple organisms have only one set of chromosomes and they are *haploid*.
2. More complex organisms have two sets of chromosomes and are *diploid*.
3. Haploid organisms have one of each kind of chromosome and one of every kind of gene.
4. Diploid organisms have two of each kind of chromosome and two of every kind of gene.
5. The two chromosomes of each kind are called **homologous chromosomes** because they are carrying the same kind of traits (genes). *Homo* means "same."
6. A human has 23 different kinds of chromosomes that are given numbers from 1 to 23. Because we are diploid organisms we have two of each of the different kinds. So, we have 46 chromosomes in all, made up of 23 *homologous pairs*.

? QUESTION

1. How many sets of DNA molecules or chromosomes does a diploid organism have? _____

2. How many sets of DNA molecules or chromosomes does a haploid organism have? _____

3. Humans are _____. (haploid or diploid)

4. How many homologous pairs of chromosomes does a human have? _____

5. Because chromosomes occur in pairs in a diploid organism, how many genes for one trait would a diploid organism possess? _____

6. How many genes for a trait would a haploid organism possess? _____ Why?

ALLELES: THE VARIOUS FORMS OF A GENE

Humans are diploid, and they have two copies of every kind of gene. One of the purposes of genetics is to figure out which form (variation) of these two genes you have, and what expression of those genes you can expect.

The *alternate forms* of a particular gene are called **alleles**. For example, there are three alternate forms—three alleles—for blood type: A, B, and O.

The reason all species have various alleles (forms of genes) is that **mutation** events change the structure of genes.

A gene can be mutated (changed) by radiation, chemicals in the environment, or other spontaneous events that are surprisingly common on this planet. There may have been a time when all the genes for eye color were identical and resulted in brown eyes. But over time, mutations occurred and changed the DNA of this eye color gene, creating a new "allele" (variation) for the eye color trait. Perhaps this new allele was for blue eyes.

Alleles are always for the same trait, and are located at the exact *same* spot on homologous chromosomes. (This is how we know that they are truly alleles of each other, and not different genes.) **Remember:** Alleles are variations of the same gene!

? QUESTION

1. What is an allele?

2. Where are alleles located?

3. What process creates the various alleles in a species? Explain how.

4. Which of the following genes (1 through 9) are alleles?

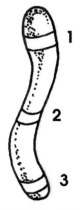

Chromosome #3

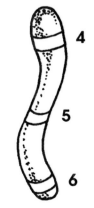

Chromosome #7

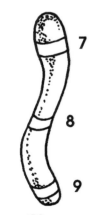

Another Chromosome #3

GENOTYPE

A *genotype* is the description of the alleles an individual possesses for a particular trait. Observe the following situation where there are two different alleles for a particular trait.

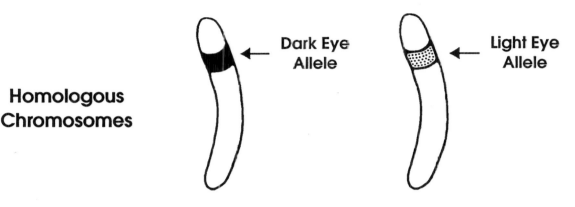

Homologous Chromosomes

Dark Eye Allele

Light Eye Allele

Note: Even though the two genes look different, they are alleles because they are at the *same locus* on homologous chromosomes.

NOW

1. Study the chromosomes above. Draw and label the three combinations of eye color alleles that are possible in individuals of the same species.

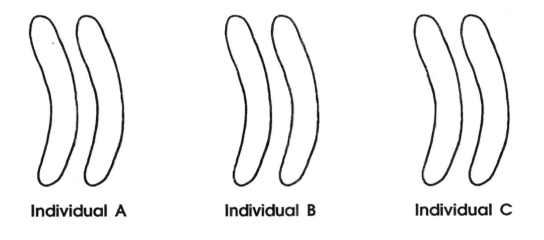

Individual A   Individual B   Individual C

_____   _____   _____

2. If an organism has *two identical alleles,* we say that it is *homozygous* for that trait (meaning the "same" two alleles).

3. If an organism has *two different alleles,* we say that it is *heterozygous* for that trait (meaning "different" alleles).

4. Go back to the diagram of the three individuals above and label each as to whether it is *homozygous* or *heterozygous.*

PHENOTYPE

The physical expression of the alleles—what an organism looks like—is termed the *phenotype*. Because there are different possible combinations of alleles (genotypes), there are alternative possible phenotypes for a trait that can be expressed in a population.

NOW

Draw this survey chart on the chalkboard, and record your phenotype for each of the six traits. After everyone in the class has recorded their phenotypes, write the class totals on the chart below. Your instructor will tell you the genotypes of these traits at the end of the next Activity.

Trait	Class Phenotype Totals		
Eye Color	*Dark # _____	Light # _____	
Earlobes	Attached # _____	Unattached # _____	
PTC Paper	Can taste # _____	Cannot taste # _____	
Hairline	Widow's peak # _____	Straight across forehead # _____	
Hair Type	Straight # _____	Wavy # _____	Curly # _____
Fingers	Five # _____	Six # _____	
Little Finger	Bent # _____	Straight # _____	
Tongue	Roller # _____	Non-Roller # _____	
Long Palmer Muscle	Present # _____	Absent # _____	

* Dark is considered to be black, brown, hazel, green, or grey.

INFORMATION

1. The phenotype is the description of the physical expression of a trait (brown eyes), whereas the genotype is the description of the exact combination of alleles (for example, 1 allele for brown eyes + 1 allele for blue eyes).

2. The genotype results from the combination of genes you inherited from your parents.

3. The phenotype results from the expression of the genes in the genotype, and also may be influenced by the organism's environment. In some cases there may be only two phenotypes for a trait, and in other cases there are *more than two* phenotypes for a trait.

ACTIVITY #2

"SOME RULES OF GENETICS"

Nine times out of ten, in the arts as in life, there is actually nothing to be discovered; there is only error to be exposed.

—H. L. Mencken
American editor and critic (1880-1956)

RULE OF THE GENE

The parent must possess the gene in order to pass it on.

The source of all genes in the offspring is the parents. Always look to the parents to figure out what genes the sperm or egg can possibly carry, and remember that a parent does not possess all of the genes found within a reproducing population of a species.

? QUESTION

1. How many different *alleles* for a single trait can a homozygous parent pass on? _____

2. How many different *alleles* for a single trait can a heterozygous parent pass on? _____

RULE OF SEGREGATION

Only one gene of the two alleles that you have is put into each gamete that you make.

Alleles are located on homologous chromosomes, and since homologous chromosomes are segregated during meiosis, the genes are also segregated.

Numerous gametes are formed during gamete production, and if the alleles are different (heterozygous), 50% of the gametes will carry one gene and 50% of the gametes will carry the other.

When alleles are the same (homozygous), 100% of the gametes will carry the same allele.

1. A parent possesses two copies of each gene. When this parent passes on its alleles for a gene, how many does it contribute to each of the offspring? _____

2. How many copies of a gene does the other parent contribute to each offspring? _____

3. How many copies of each gene for the trait does each offspring receive? _____

RULE OF DOMINANT AND RECESSIVE ALLELES

Some alleles control the phenotype even if they are paired with a different allele.

An obvious example is the dark eye allele that will create the dark eye phenotype in an individual even if the allele for light eyes is present.

If two different alleles are together in an organism, and only one phenotype is expressed, then the allele that is expressed is called *dominant*.

The other allele that is "hidden" is called *recessive*.

? QUESTION

1. Can the individual carry an allele that is not expressed? Explain.

2. What word is used to describe the *genotype* condition where there are two different alleles together in the same organism? _____

3. What word is used to describe the *genotype* condition where there are two of the same alleles together in the same organism? _____

INFORMATION

Since dominance and recessiveness have intricate biochemical explanations, the only way of determining dominance is to cross two individuals that are homozygous (pure) for the two different phenotypes. This produces the heterozygous condition. Whichever phenotype is exclusively expressed is said to be the *dominant phenotype*.

1. A homozygous blue-eyed mouse with short whiskers mates with a homozygous brown-eyed mouse with long whiskers. All of their offspring have brown eyes and short whiskers. Which alleles are dominant?

2. A homozygous five-clawed cat is crossed with a homozygous six-clawed cat and all of the kittens have six claws. Which allele is dominant?

3. In humans, the five-fingered condition is *recessive* to the six-fingered condition. Yet, most people have five fingers. Explain how this can happen.

IMPORTANT MESSAGE

Ask your instructor which alleles are dominant in the class Phenotype Chart. It is a common *mistake* to assume that the allele found most frequently is always the dominant allele. *Natural selection* determines the success of an allele.

RULE OF INCOMPLETE DOMINANCE

When two different pure-breeding strains are crossed, and their offspring show a blending of phenotypes, then neither allele is dominant.

This is easily recognized when the phenotype is somewhere between two extremes. Counting the parents, there are *three* phenotypes (*black, white, grey*) being expressed in these flowers instead of only two, and that third phenotype is *intermediate* between the other two. This heterozygous condition is called **incomplete dominance**.

1. On the chart you did earlier, which of the three hair types (wavy, curly, or straight) represents incomplete dominance—the *blended* heterozygous condition? _____

2. You cross a herd of red cattle with white cattle and all of the calves appear to be roan (reddish white). Is this an example of incomplete dominance? _____ How do you know?

3. You cross a blue flowering pea plant with a white flowering pea plant and all of the offspring are blue flowered. Is this an example of incomplete dominance? _____ How do you know?

ACTIVITY #3

"HOW TO SOLVE GENETIC PROBLEMS"

USING LETTERS FOR ALLELES

For convenience, the genes of an allele pair are usually symbolized by a letter from the alphabet. A *large* letter is used for the dominant trait and a *small* letter for the recessive trait. When we want to describe the *genotype* of an organism, we use both letters to represent the alleles inherited from the parents.

For example, free earlobes is a dominant allele and attached earlobes is recessive. You would use a capital "**F**" to indicate the dominant allele and a small "**f**" to indicate the recessive allele in describing an individual.

? QUESTION

1. Write the three genotypes for earlobe attachment as it applies to the following individuals.

    a. Heterozygous _____ _____

    b. Homozygous Dominant _____ _____

    c. Homozygous Recessive _____ _____

2. When it comes to symbolizing incomplete dominance with letters, it is best to use the letter "**C**" for one allele and "**C'**" for the other allele.

    List the three possible genotypes for hair type.

    a. Curly _____ _____

    b. Wavy _____ _____   Why not use a small letter "**c**" for the heterozygous genotype?

    c. Straight _____ _____

USING THE PUNNETT SQUARE

The **Punnett Square** is a method of predicting the probable outcome of genetic crosses.

STEP 1   Draw a square like this:

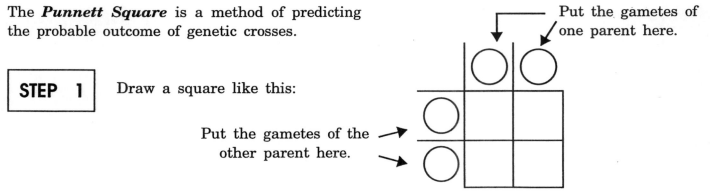

Put the gametes of one parent here.

Put the gametes of the other parent here.

Determine what kinds of gametes are made by each parent in the cross, and put those gametes into the boxes of the Punnett Square.

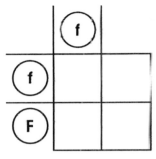

If your mate is heterozygous (Ff), then those gametes are F and f.

For example, if you are homozygous recessive for attached earlobes (ff), then all of your gametes are f.

STEP 3   Fill in the offspring boxes of the Punnett Square.

In this example there are only two possible offspring genotypes. The Punnett Square tells us to expect about 50% ff and 50% Ff.

Sometimes the Punnett Square is more complex than this and you must figure out more than one trait at a time. Nevertheless, you use the same basic method.

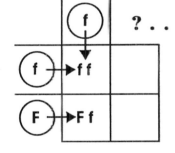

? . . . Why is it unnecessary to fill in this box with another small f?

NOW

Make up your own genotype example and work out the crosses.

1. Traits:

2. Symbols:

3. Male Genotype:

4. Female Genotype:

5. Offspring Genotypes:

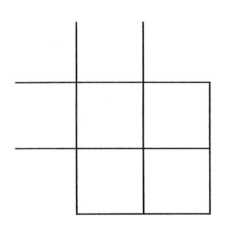

Punnett Square

ACTITITY #4

"GENETIC PROBLEMS"

CASES OF COMPLETE DOMINANCE

1. Gregor Mendel grew different varieties of pea plants in his garden. When he crossed yellow-seed plants with green-seed plants, he always got yellow pea seeds.

   a. What is the dominant allele?

   b. What is the genotype of all green-seed plants?

   c. Use the Punnett Square to show Mendel's cross.

   d. Do the parent yellow-seed plants have the same genotype as the offspring yellow-seed pea plant?

   Parent: _____ Offspring: _____

   e. What genetic fact do you know about any yellow-seed pea plant?

   f. If yellow-seed pea plants are dominant to green-seed pea plants, why are there mostly green pea seeds in nature?

2. A dark-eyed man mates with a light-eyed woman and they have ten dark-eyed children.

   a. What is the dominant allele?

   b. What is the genotype of all light-eyed people?

   _____

   c. What are the genotypes of the two parents?

   _____ and _____

   d. What is the genotype difference between the dark-eyed parent and the dark-eyed offspring?

   Parent: _____ Offspring: _____

   e. When two heterozygous dark-eyed people (Dd) are crossed, what is the *phenotype* ratio of dark-eyed offspring to light-eyed offspring? (Use the Punnett Square to get your answer.)

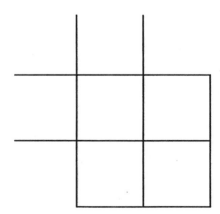

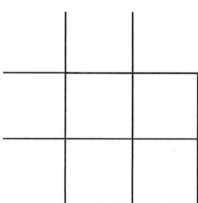

A SPECIAL NOTE ON EYE COLOR

Eye color is probably due to multiple alleles and more than one gene pair. The numerous phenotypes are determined by genes that control both the *amount* and the *distribution* of a dark pigment called **melanin**. Except for albinos, everyone has some eye pigmentation.

Eye color is determined mainly by the location of melanin in the iris of the eye. Concentrated melanin particles appear as brown; dilute melanin particles appear as yellow or yellow-brown.

VARIOUS EYE COLORS

Blue: no melanin in the front part of the iris. The color is due to minimal amounts of melanin in the rear of the iris with the clear front portion scattering the light reflected off the melanin. This scattering is greatest in the blue spectrum giving the iris its blue color.

Grey: the same as blue, but with a slight amount of melanin in the front of the iris which tones down, or greys, the blue reflected from behind.

Green: a bit more melanin particles scattered in the front part of the iris create yellow. Blended with the light blue from the rear of the iris, it produces an overall green color.

Hazel: even more melanin particles in the front of the iris give a slight brown color, and dilute melanin particles scattered throughout the iris add some yellow.

Brown: melanin particles in the front part of the iris and throughout the iris. The amount of melanin varies, leading to gradations of brown color in the eye.

Black: large amounts of melanin in the front and throughout the iris.

TEST CROSS TO CHECK GENOTYPE

If an organism shows the dominant phenotype, then one of its genes has to be the dominant allele, but you cannot be sure of the identity of the other allele unless you do a **test cross** to see if the dominant parent breeds pure. Let's pretend that you are in the dog-breeding business. You know that long hair on a "pooch hound" is a dominant allele and short hair is recessive. You purchase a male long-haired "pooch hound." How do you figure out if your male "pooch hound" is homozygous or heterozygous for long hair? Which genotype of female should you breed him to?

If a proper test cross is used, what phenotypes of puppies would you see if your male dog is heterozygous dominant? _____
What puppy phenotypes would you see if your male dog is homozygous dominant? _____ Complete the Punnett Square to show the test cross that would convince someone that your "pooch hound" is homozygous for long hair.

    Male Genotype: _____ Female Genotype: _____

CASES OF INCOMPLETE DOMINANCE

1. When a straight-haired mouse is crossed with a curly-haired mouse, the result is always wavy hair. Two wavy-haired mice cross.

   a. What are the genotypes of the two wavy-haired mice? _____

   b. Draw the Punnett Square of a cross between two wavy-haired mice, and show the probable genotypes of their offspring.

   c. What is the expected *phenotype* ratio of the offspring?

   _____ % _____ % _____ %

   d. What is the expected *genotype* ratio of the offspring?

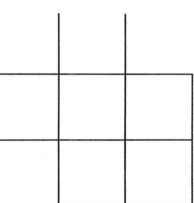

2. Red orchids with straight petals are crossed with white orchids with curly petals. The results are pink orchids with wavy petals.

   a. What are the genotypes of the two parent orchid plants? *Remember:* You are dealing with *two different* traits.

   First parent: ___ ___   ___ ___
                 (color)   (shape)

   Second parent: ___ ___   ___ ___

   b. What is the genotype of the offspring orchids?

   Offspring: ___ ___   ___ ___

ACTIVITY #5

"SEX-LINKED TRAITS"

SEX DETERMINATION

Humans have 23 homologous pairs of chromosomes. Twenty-two of these pairs are named using the numbers 1 through 22. The 23rd pair is individually labeled with the letters "**X**" and "**Y**" for males, and "**X**" and "**X**" for females. These labels distinguish them as the **sex chromosomes**.

During meiosis in the male two types of sperm are produced: those carrying the X and those carrying the Y chromosome. Females produce eggs carrying only the X chromosome.

If a Y chromosome is present in the cells of an embryo, then the child becomes a male. If the Y is not present, the child becomes a female. It is the presence or absence of the Y chromosome that determines the sex of a child!

This means that a male child receives a Y chromosome from his father and an X chromosome from his mother. A female child receives an X chromosome from her father and the other X chromosome from her mother.

Y + X = Male   X + X = Female

? QUESTION

Draw a Punnett Square to show a cross of X and Y chromosomes in the fertilization of male and female gametes.

The offspring boxes should reveal why we have about a 50% male to 50% female ratio within the human population.

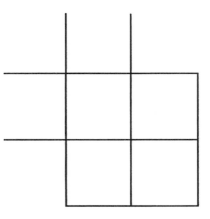

SEX-LINKAGE

The X and Y chromosomes are not exactly identical, and we should expect that there would be differences in how each of them carries genes. These differences are expressed in the unequal frequencies of phenotypes in the male and female offspring.

If any phenotype is distributed *unequally* between male and female offspring, and those differences are due to X and Y chromosome differences, then we call those traits ***sex-linked***. Actually, "sex-linked" means that the gene is carried on the X chromosome and not on the Y chromosome. We would call these genes ***X-linked***. It is easier to understand sex-linkage by looking at the sex chromosomes.

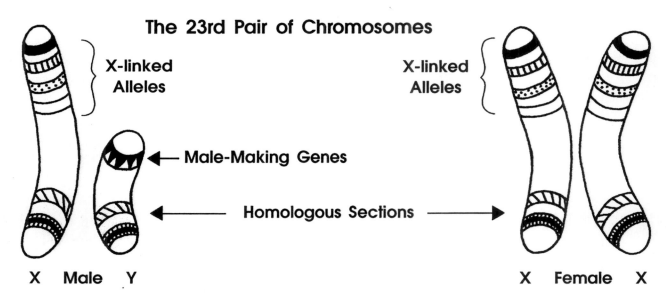

The 23rd Pair of Chromosomes

Two genetic situations are illustrated above.

First: There is a homologous section of the X and Y chromosomes that is the same, and there will be no differences in phenotype between male and female children.

Second: Notice that the Y chromosome is very short. We would expect it to lack some of the genes that are carried on the X chromosome. There is an X-linked section on the X chromosome that carries genes that are *missing* from the Y chromosome.

? QUESTION

1. How many copies of an X-linked gene does a male have? _____

2. Will a male be able to give X-linked genes to his daughter? _____ Explain.

3. Will a male be able to give X-linked genes to his sons? _____ Why or Why not?

226

4. How many copies of an X-linked gene does a female have? _____

5. A male child gets X-linked genes from which of his parents? _____

6. A female child gets X-linked genes from which of her parents? _____

7. If a father is carrying an X-linked allele, then how many of his sons will get that allele? _____

   How many of his daughters will get that allele? _____

8. If a mother has a defective X-linked allele on one of her chromosomes and the other chromosome is normal, then how many of her sons will get that defective allele? _____

   Will any of her daughters get the defective allele? _____ How many? _____

9. If we found that none of the daughters actually showed the defective phenotype, then how could we explain it?

TIPS FOR SOLVING SEX-LINKED GENETIC PROBLEMS

There is a sex-linked gene on the X chromosome that causes a disorder called *hemophilia*, where the blood fails to clot properly when a person is injured. This disorder is recessive and can be symbolized by the small letter "n." Normal blood clotting is dominant and can be symbolized by the capital letter "N."

In sex-linked cases we not only use letters to symbolize the genes, but also include the X or Y chromosome to indicate gender and to follow the sex chromosomes into the next generation.

Using these symbols we would indicate a female who is heterozygous for clotting as $X^N X^n$.

A homozygous female for normal clotting would be $X^N X^N$.

A hemophilic male would be $X^n Y$.

We would diagram a Punnett Square of a cross between a heterozygous female and a normal clotting male like this:

Complete the Punnett Square showing the offspring.

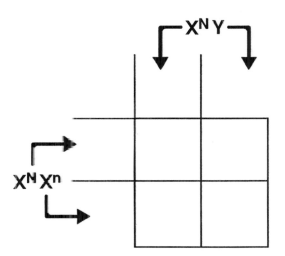

1. Looking at the Punnett Square you just completed, answer the following questions.

    a. What is the genotype for the female parent? _____

    b. What is the genotype for the male parent? _____

    c. What are the genotypes for their offspring? _____

    d. What are the chances that any child will be a hemophiliac? _____

    e. Is it the father or the mother that passes the hemophilia gene to the male child?

    _____

2. Failure to distinguish between red and green colors is a recessive allele and is a sex-linked gene carried on the X chromosome.

A red-green color-blind male mates with a normal female. Of their six children (four boys and two girls), all have normal vision.

    a. What is the most probable genotype of the mother? _____

    b. Will any of the male children pass this disorder·on? _____ Explain.

    c. Draw a Punnett Square of this cross to prove your answers.

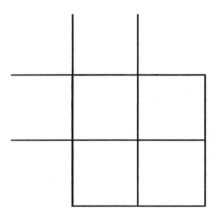

3. A normal-visioned female gave birth to a color-blind daughter. Her husband has normal vision. He claims that the child is not his. Does the genetic information indicate that someone else is the child's father?

Explain and prove your answer using a Punnett Square.

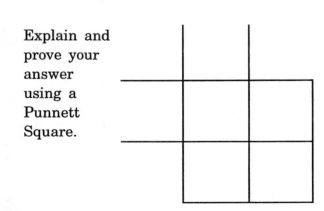

Green Apples Only

Red Apples Only

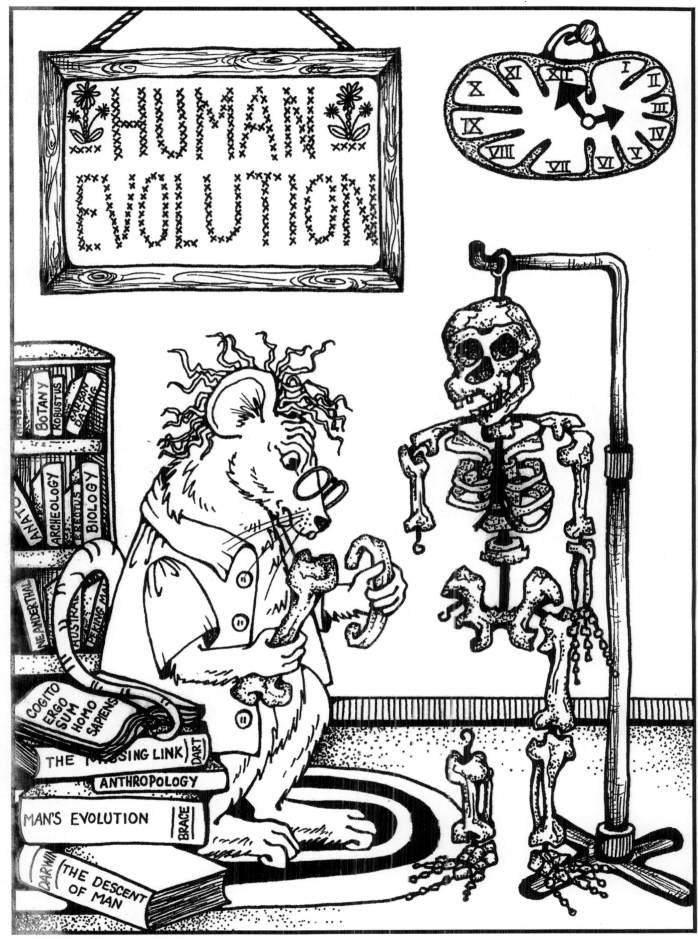

229

Summary Questions

1. What is meant by the "Mitochondrial DNA Clock," and how is it used to compare ancestral groups of humans?

2. What happens to the DNA of two groups of humans once they become reproductively isolated from each other?

3. What happens to the DNA of one group of humans that remains geographically isolated from all other groups for a long period of time?

4. Where did humans originate on this planet? Discuss the evidence you have to support your answer.

5. Define, and describe the differences between *Australopithecus*, *Homo habilis*, and *Homo erectus*.

HUMAN EVOLUTION

INTRODUCTION

People are very curious about where things come from, and what they mean. This applies especially to human history. The many written stories about humans are all we had to tell us about our past, until anthropologists discovered that we could learn more about humans before written history.

Now, we are able to present the scientific story of man's evolution on this planet using fossil record and DNA evidence. All that you will need to join in this discussion is practice with archeological and biological information. During this lab you will learn to develop a scientific history of humans, using several techniques of thinking, map-making, and group interaction.

ACTIVITIES

ACTIVITY #1

"TRAILS OF THE CROSS-COUNTRY HIKERS"

This first problem will develop your thinking method for understanding a scientific history of modern humans.

The information provided may seem to be brief. But, as you think through the problem, you will be able to draw a map of the different paths taken by the five students, indicating when they separated from each other, and the trails they took.

Your time map will look something like this, using branches to represent when the different hikers separated from each other.

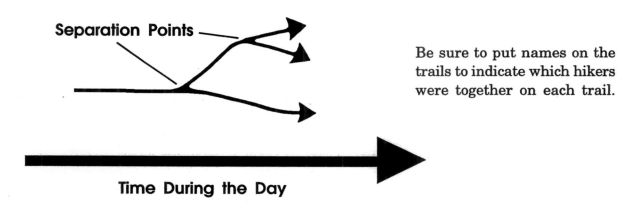

Separation Points

Be sure to put names on the trails to indicate which hikers were together on each trail.

Time During the Day

NOW

1. Divide into groups of 3 or 4 people.

2. Your group may need some quiet space to do this Activity, so don't hesitate to work outside or in another room if necessary.

GO GET

1. Refer to the worksheet at the end of this lab titled, "Trails of the Cross-Country Hikers."

2. A five-card set of names. You can use these name cards as an aid to physically keep track of the hikers during your analysis of the problem.

"Cross Country Hikers"

Five members of the Cross-Country Running Team decided to have a rugged day of fun last Saturday. They drove out a long dirt road, and parked the car at a grove of trees that was 12 hours of hard hiking away from the freeway. All they had to do was to walk west and they were guaranteed to reach the Highway 8 Freeway, where they could hitch a ride back to their parked car.

Your task is to figure out the general paths that were taken by the individual people using the limited information provided below, and to draw those paths on a map.

INFORMATION

1. The five hikers started out together, but divided into smaller groups, taking different paths as they went along.

2. Below you will find clues about the different people. The clues tell you when particular hikers separated from each other and took different paths. You will have to use those clues to figure out and draw their approximate paths to the highway.

3. The hikers names are: Bill, Hector, Julie, Tom, and Maria.

4. The hikers all began to walk at about 6:00 a.m. All of the hikers reached Highway 8 at around 6:00 p.m., but they did not necessarily arrive together.

CLUES

A. Tom and Maria arrived at Highway 8 together.

B. The last time Hector was with Julie was 5 hours before he reached the highway. (This clue does not tell you whether or not Hector or Julie were with anyone else when they separated from each other. You will have to figure that out—read on!)

C. The last time Julie was with Tom was 8 hours before she arrived at the highway. (Again, this clue does not tell you whether or not Julie or Tom were with anyone else when they separated from each other.)

D. The last time Bill was with Hector, they were 10 hours away from the highway. (Was Hector or Bill with anyone else at the time they last saw each other? Again, this clue does not tell you. Work it out!)

233

1. Draw a map of the trails taken by these five hikers. Show when they split up during the day, and how they were grouped when they arrived at the highway at 6:00 p.m.

2. Make a practice map to work out the problem, then draw your final map on the "TRAILS" worksheet at the end of this lab.

3. You may find it helpful to use the name cards to keep track of the hikers. Lay them on your paper, on the ground, or on the table, and move them along the trails as you work through the clues.

FINISHED

1. Check with your instructor to make sure you got the correct answer to this problem before going on.

RED FLAG!   WARNING!   WARNING!

2. The next problem is going to take much longer, and will use the same kind of thinking you used for the Cross-Country Hiker problem.

3. *Good advice:* Have the best reader in your group read aloud as the rest of your group reads Activity #2, "The Mitochondrial DNA Clock."

Work together.

Work slowly.

When you have questions, check with your instructor.

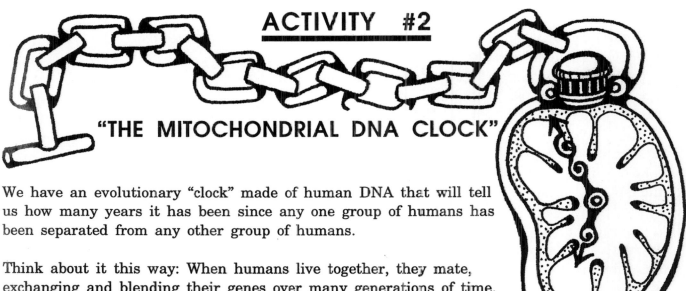

ACTIVITY #2

"THE MITOCHONDRIAL DNA CLOCK"

We have an evolutionary "clock" made of human DNA that will tell us how many years it has been since any one group of humans has been separated from any other group of humans.

Think about it this way: When humans live together, they mate, exchanging and blending their genes over many generations of time. This is how sexual reproduction creates common traits in human DNA.

Now, what happens if one group splits into two separate groups that migrate far away from each other, and they never get a chance to interbreed with each other again?

As you would expect, the two groups are no longer mixing genes (and DNA), and over time they will begin to look somewhat different from each other.

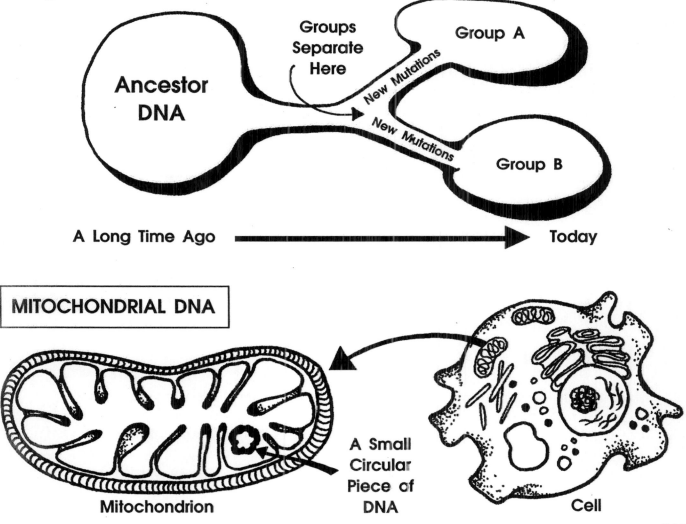

235

Humans have DNA in the mitochondria. This mitochondrial DNA does not mix with the DNA in the cell nucleus during sexual reproduction. So, the only way it gets passed onto the next generation is through the mitochondria in the *mother's egg*. The father's sperm does not give mitochondria to the egg during fertilization. This strange situation means that mitochondrial DNA (passed on by the mother) stays the same, and cannot change except by "new" mutations. These mutations occur randomly as a natural process of living on this planet.

By comparing DNA patterns, chemists can detect any new mutations that have been added to the small piece of original mitochondrial DNA that was passed on from mother to child through thousands of human generations.

If a group of humans splits off from a common ancestral group, then both groups will begin to accumulate different mutations from each other. Each group is genetically separated from the other group. This starts at the point where they no longer interbreed.

The easiest way to show that two groups have separated in the past is to count the number of *new* mutations found in the DNA of one group and *not* found in the DNA of the other group.

The *greater* the number of *different* mutations, the longer is the amount of time that the two groups have been separated from each other. (**Remember:** Mutations take a lot of time to happen.) If however, the DNA of the two groups is quite similar, then they have not been separated very long.

COMPARATIVE DNA ANALYSIS

Diagrammed below is a comparison of the mitochondrial DNA of three geographically separate human racial groups. Each **bar** represents a mutation.

? QUESTION

1. Which two groups are the most similar?

2. Which one of the three groups is the most different from the other two groups?

3. So, which group has been separated from the others for the longest time?

4. And, which two groups haven't been separated from each other for very long?

THE "CLOCK"

Biologists have studied many different species, including humans, and have estimated that it takes 500,000 years for 1% of the mitochondrial DNA to be changed by mutation. (This estimated mutation rate is currently being debated, but we will base our calculations on it. Perhaps more research will clarify our understanding of evolutionary timelines.)

This mutation rate is a kind of "clock."

Researchers have gone around the world collecting mitochondrial DNA samples. They have counted the total number of small mutations that have occurred in the humans of today. Each one of those mutations represents an amount of "time" on the mitochondrial clock.

? QUESTION

A representative sample of different human geographical/racial DNA has been collected and analyzed. When all of the different mutations were counted, scientists found that only 0.4% of the DNA of modern humans has been mutated. How many years have *modern* humans been on this planet?

Remember: It takes 500,000 years for 1% of the DNA to be mutated.

? QUESTION

Can you estimate how long geographical/racial groups have been separated from each other?

HOW TO DO IT

If you want to estimate how long it has been since two racial or geographical groups have been separated, you must:

1. Use the figure for the % of genetic difference between the two groups (as determined by genetic researchers).

2. Multiply the % of genetic difference by 500,000 years. (It takes 500,000 years for 1% of the mitochondrial DNA to be mutated.)

3. *Practice Example:* If Group A and Group B have 0.1% genetic difference, then it has been about _____ years since they separated, and have not had the opportunity to interbreed.

500,000 years x 0.1% = 50,000 years

Answer: 50,000 years

Calculate the answer to the following problem:

Group A, the "Altitudinals," have lived in the high mountain ranges of Nepal for longer than anyone can remember. Group B, the "Basinals," have lived on the flat river delta plains of Southern India for hundreds of centuries according to their written history.

A genetic researcher has measured a 0.05% genetic difference in the mitochondrial DNA of Group A and Group B.

How long would you say it has been since these two groups lived together and had the opportunity to interbreed? In other words, how long ago was it that the Altitudinals and the Basinals split company, one moving to the highlands and the other moving to the lowlands?

_____ years

Show your work here:

FINISHED

1. It's time to move on to Activity #3, the "Time Trails of Modern Humans," and test your understanding of the "Mitochondrial Clock."

2. If you are having trouble understanding the mitochondrial clock, then *now* is the time to get help from your instructor!

ACTIVITY #3

"TIME TRAILS OF THE MODERN HUMANS"

Archaeologists have found skeletal evidence of a modern-type human (looks like people today) who lived on this planet about 100,000 years ago. There is a question as to exactly how long modern humans have been here, and where they might have originated.

You will investigate a partial answer to these questions. Modern human types are known as *Homo sapiens*. They are somewhat lighter in build, with a bit larger brain size than the previous human fossils. Evidence also suggests that they lived much like the hunter-gatherer peoples of today. If you were to bring some of these skeletal remains back to life, put some modern clothes on them, and put them on a bus, you could not tell them apart from anyone else on that bus.

Where did modern humans originate on this planet?

What is the "trail" that modern humans took when spreading out to the different continents on this planet?

Figuring out the answers to these questions is somewhat like doing the "Trails of the Cross-Country Hikers" problem. Here, you will follow five geographical/racial groups instead of five hikers.

You must calculate when these groups separated from each other, and relate those times of separation to a world map. From that information, you will be able to tell a story of modern human migration and origin.

HOW TO DO IT

Answer the following questions using what you learned about the Mitochondrial DNA Clock.

1. When genetic researchers compare the DNA of *Northern Asian* populations versus *Native American* populations, they find a 0.07% difference. The "Mitochondrial Clock" formula tells us that there has been about _____ years since the separation of these two groups.

2. When comparing the DNA of *Northern Asian* and *European* populations, they discover a 0.10% difference. The "Mitochondrial Clock" formula tells us that there has been about _____ years since the separation of these groups.

3. When comparing the DNA of *Indonesian* peoples to either the *European* group or the *Northern Asians*, they find a 0.12% difference. The "Mitochondrial Clock" formula tells us that there has been about _____ years since separation.

4. When researchers compare the DNA of *African* populations to any other group, they find a 0.20% difference. The "Mitochondrial Clock" formula tells us that there has been about _____ years since the separation of Africans from other groups.

GO GET

1. Refer to the "Time Trails of Modern Humans" worksheet at the end of this lab.
2. Refer to the "World Map" worksheet at the end of this lab.

NOW

Use the "years since separation" information you just calculated on the previous page for the different geographical/racial groups, and . . .

Fill out the map of the "Time Trails of Modern Humans" on your worksheet. Show when each group split off from the original group (just like you did for the Cross-Country Hikers).

THEN

Using the "World Map" worksheet and the "Time Trails" map you just completed, draw your best interpretation of the trail taken by humans as they separated and spread out on this planet. Show where the original population started and show the sequences of where they separated and where they went after that. There may be more than one possible trail based on the genetic information presented in this Activity.

FINALLY

Below is a table of information about the general anthropological evidence of the earliest settlements of modern humans that have been excavated so far.

Region	Time of the Earliest Settlements
Americas	less than 35,000 years ago
Indonesia	50,000 years ago
Europe	35,000 years ago
Asia	60,000 years ago
North Africa	100,000 years ago

How does your story, illustrated by the Migration Map, compare to this evidence?

If you think that all of this is hard to figure out, remember that many researchers have spent their careers just to give you the small amount of information you are working with in this lab. There is a great deal more to know that will make the picture more accurate. Anthropology is one of the sciences of biology. Archeology is another. If you are enjoying what you are learning in this lab, consider studying in these other fields of biology as well.

240

ACTIVITY #4

"THE STORY OF THE REALLY OLD HUMAN-LIKE PEOPLE"

Dating back to a time before *Homo sapiens* (modern humans) appeared on this earth, there are a number of very old fossils that give clues about early hominid evolution and migration.

1. Australopithecus and other early hominids

The Australopithecines are among the oldest hominid fossils to be found. They were smaller than modern humans, standing approximately four feet high, walked slightly bent over compared to modern man, and they had a smaller brain. Other skeletal differences suggest that there were perhaps a dozen species with different lineages and niches. Their fossils occur between 6 and 1.5 million years ago, and one of them may be a direct ancestor to *Homo habilis* and *Homo erectus*.

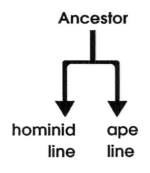

Ancestor

hominid line   ape line

2. Homo habilis

The *Homo habilis* group was a tool maker. They had a somewhat larger brain than *Australopithecus*, stood more erect, and were slightly bigger in overall skeletal size than *Australopithecus*. Compared to *Homo erectus*, who also made tools, the *Homo habilis* group had a smaller brain capacity.

3. Homo erectus

These fossil groups, designated as Peking, Java, and Heidelberg, stood fully erect (for which they are named), and had the same general height and brain size as modern man. They have been found with more artifacts than other hominids, telling us something about how they lived. Apparently, they were very much like *Homo sapiens*, but perhaps had less ability to speak because of some throat structure differences. There is evidence suggesting that *Homo erectus* may have included more subgroups than the three mentioned above, and their migration out of Africa might be a much more complex story.

| GO GET |  |

Refer to the second World Map worksheet at the end of this lab.

| NOW |

Using the information provided on the next page and the World Map worksheet titled "Trails of the Early Hominids," put a dot on the map for each of the groups. See if the dots suggest a story of how these human-like species might be related to each other, and where they developed and migrated.

241

Fossils	Age of Fossils	Where Found
Ardipithecus ramidus	5.5 million years	Africa
Australopithecus anamensis	4.2 million years	Africa
Australopithecus afarensis	3.5 million years	Africa
Australopithecus africanus	3.5 million years	Africa
Homo habilis	2.5 million years	Africa
Homo erectus	1.8 million years	Africa
Homo erectus (Peking)	500,000 years	China
Homo erectus (Java)	1.5 million years	Southeast Asia
Homo erectus (Heidelberg)	900,000 years	Germany

? QUESTION

1. Based on available fossil evidence, where did the various early hominids originate?

2. Which group migrated out of the continent of origin?

   How long ago?

   How far did that group spread over the world?

3. Was there another group that later came out of the continent of origin? (See Activity #3.)

   Which species was that group?

   When did the migration of the new species happen?

   Indicate that event on your map.

FINISHED

Show your completed map to your instructor.

HIGHWAY 8

Bill
Hector
Julie
Tom
Maria

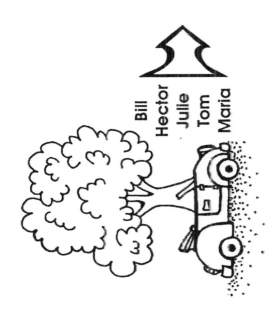

| 6:00 | 8:00 | 10:00 | 12:00 | 2:00 | 4:00 | 6:00 |
| A.M. | | | NOON | | | P.M. |

" TIME TRAILS of MODERN HUMANS "

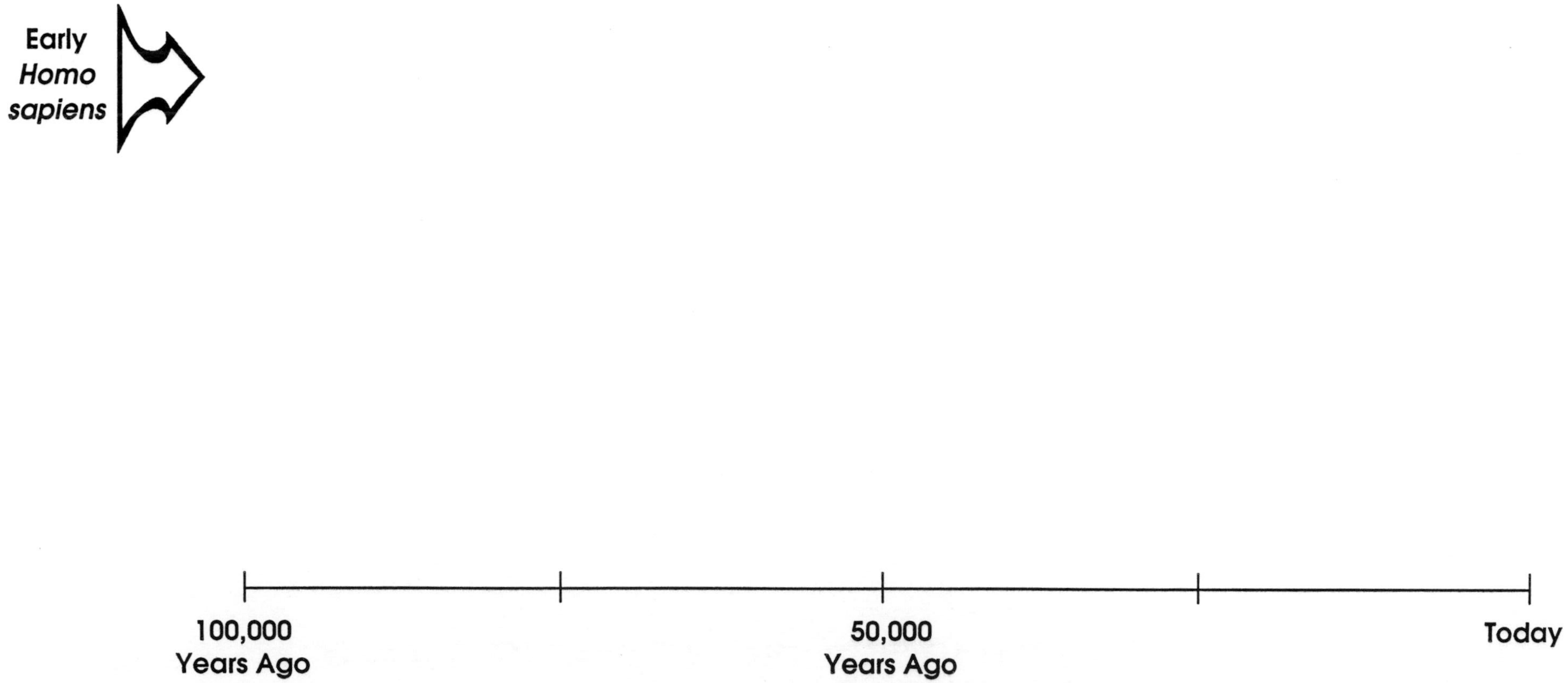

Early
*Homo
sapiens*

100,000
Years Ago

50,000
Years Ago

Today

WORLD MAP

" Migrations of Modern Humans "

WORLD MAP

" Trails of the Early Hominids "

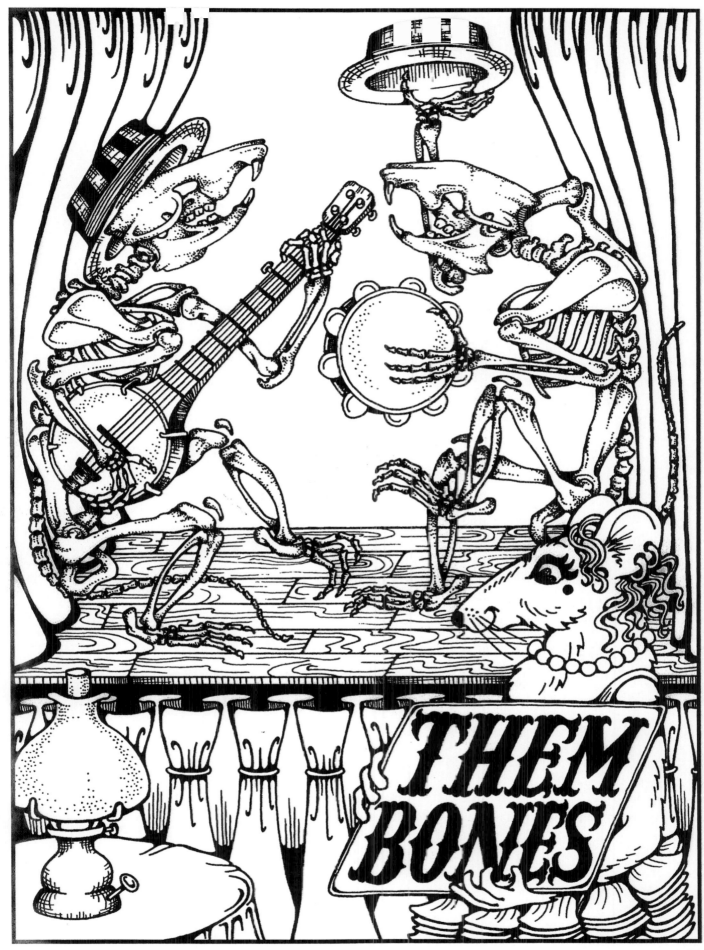

Summary Questions

1. List four important functions of your skeletal system.

2. Give examples of how the matrix and density of cells differ in the various connective tissues.

3. What is the difference between a tendon and a ligament?

4. Draw a very simple cross-section of a bone showing where the yellow and red marrow are located.

5. What is red marrow and what does it do?

6. Discuss how an injury may effect the second cervical vertebrae in a way that causes death, paralysis, or serious whiplash.

7. What are the three parts of the pelvis, and where do you feel each part?

8. Discuss the most obvious differences between the fetal skeleton and the adult skeleton (other than size).

THEM BONES

The skeletal system is a collection of bone and connective tissues. It provides support and protection for various organs, and operates as a lever system for the muscular movement of the body. In addition, bone marrow is the source of all red blood cells and most white blood cells.

This week's lab presents a microscopic view of bone and the connective tissue in skin, tendons, and ligaments. Also, you will be learning the names of some bones and various features of these bones.

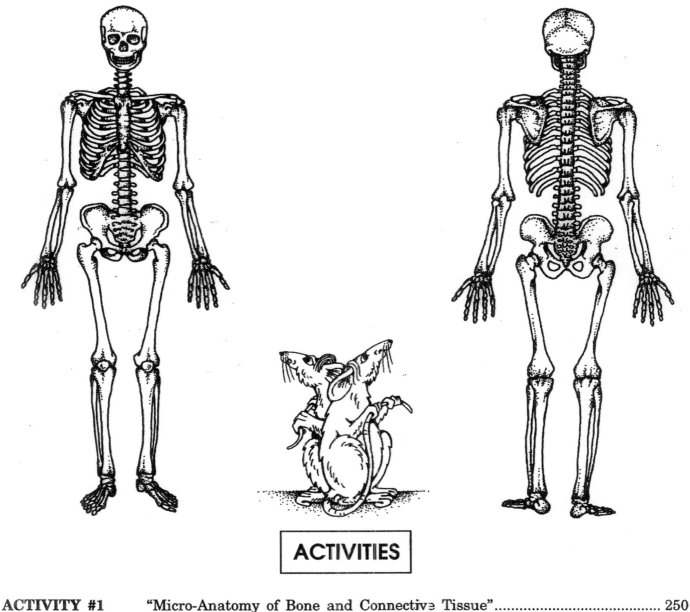

ACTIVITIES

ACTIVITY #1

"MICRO-ANATOMY OF BONE AND CONNECTIVE TISSUE"

Bone and connective tissues support the functional form of body parts. These tissues consist of specialized cells and a **matrix** of various minerals and fibers produced by those cells. The characteristics of the matrix determine whether the tissue is rock-hard (like bone), very elastic (like the skin and blood vessels), or intermediate in rigidity (like tendons and ligaments).

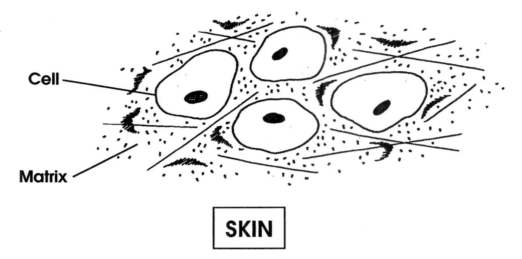

Cell

Matrix

SKIN

Skin is very elastic because of the stretchy fibers in its connective tissue.

GO GET

Go to the demonstration microscope marked "skin."

NOW

1. Find the connective tissue fibers.

2. Draw a simple sketch of the tissue showing the arrangement of fibers. Show the *orientation* (parallel, crisscross, or whatever they are) and the *variety* of fibers.

Skin

250

1. Were you able to find two kinds of fibers? If not, go back and look for long, thin, string-like fibers and thicker, very irregularly-shaped fibers.

2. How is the orientation of the long thin fibers related to your skin's ability to stretch in all directions?

3. As your skin stretches many times in a particular way (smiling) during the years, wrinkles begin to form. What do you think happens to the orientation of the fibers as you age?

4. Is a $100 face-cream going to change the orientation of these fibers?

TENDONS AND LIGAMENTS

A *tendon* is a sturdy fibrous cord that attaches the muscle to a bone. Tendons move quite a lot as your muscles contract and relax.

A *ligament* is made of very strong fibrous bands that attach one bone to another. Ligaments allow only enough movement for the joint to flex.

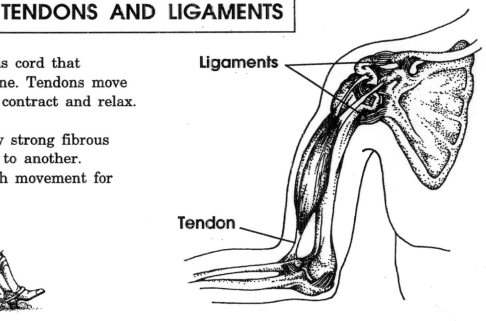

Ligaments

Tendon

GO GET

Go to the demonstration microscope marked "tendon."

NOW

1. Notice the difference in the matrix of the tendon tissue compared to skin.

2. Draw a simple sketch of the tendon tissue, indicating the relative density of cells compared to the matrix. *Hint:* The dark-stained structures are the cells.

Tendon

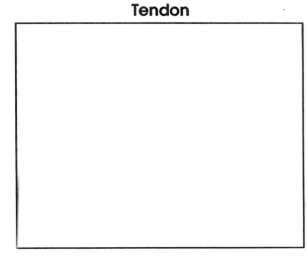

1. Which fibers are found in the skin, but are not found in the tendons?

   What does that tell you about tendons?

2. A round open hole in the tissue is a blood vessel. Do you see any? (Include these in your drawing as a reminder that they are there.)

3. Tendons and ligaments heal very slowly after injury. How does this relate to the density of cells in this tissue?

BONE

Bone is a very hard connective tissue because the matrix is composed of mineral salts (mainly calcium phosphate and calcium carbonate). A typical bone (such as the leg) begins as a cartilage structure before birth, but that growing cartilage structure is gradually replaced by bone during childhood.

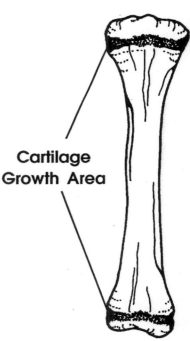

Cartilage
Growth Area

GO GET

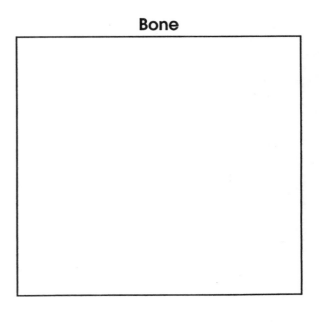

1. A compound microscope.

2. A microscope slide of bone.

NOW

1. Look at the bone section under low magnification at first.

2. You should be able to see many round or oval dark areas surrounded by concentric rings of smaller dark cavities all connected by fine canals. These smaller cavities contain **bone cells**. The larger cavity in the center is where a *blood vessel* passes through the bone.

3. Make a sketch of the arrangement of the cavities and canals in the bone tissue.

Bone

1. How many cells are in the bone tissue compared to the number of cells in tendons?

2. How many blood vessels are in bone tissue compared to tendons?

3. Bone heals quickly after an injury. How does this relate to your answers above?

NOW

1. Go to the demonstration table and examine the bone that has been cut in half lengthwise.

2. You should be able to see several distinct features.

   First, the large open chamber in the bone is called the ***marrow cavity***, and it is filled with yellow marrow (fat).

   Second, the inside part of the ends of the bone are more porous than the outside of the bone. This spongy bone material is called ***red marrow*** because it is where red blood cells are produced. Most white blood cells are also produced here.

   Third, the outside bone is compact and provides strength. It is the area that you saw in the bone slide.

3. Draw a sketch of the demonstration bone showing all of the above three features. Label them.

Lengthwise Section of Bone

ACTIVITY #2

"THE SKELETON"

A general examination of the skeleton would require weeks of study, so we will cover only some of its features during this Activity.

SKULL

The human skull consists of 28 bones that are mostly fused. These bones form a protective chamber for your brain as well as making a chewing jaw. The skull also provides locations for your eyes, ears, and nose.

GO GET

1. A human skull. (*Please handle the skull carefully.*)

2. A long bristle hair (for probing into holes).

NOW

Find the following structures and show them to your instructor:

1. **Optic Orbits:** The common term is eye sockets. If you look inside and probe with the hair, you should see several passageways into the brain. These canals and fissures are for blood vessels and nerves.

2. **Small Foramen:** These small canals are found throughout the skeletal system. You should be able to locate at least six on the front of the face. Find them. These passageways are for nerves and blood vessels.

3. **Foramen Magnum:** Turn the skull upside-down and find the large opening where the spinal cord enters the brain. If a head injury causes swelling in the brain, the base of the brain is forced into the foramen magnum. Your breathing and blood pressure centers are located in the base of the brain. This is one reason that a person with a head injury can quickly die.

4. **Zygomatic Arch:** The common name for this is the cheekbone. Is it solid? Put your thumb up under your zygomatic arch while you bite down hard. You should be able to feel a muscle contract. Now you know that a muscle passes under the zygomatic arch and attaches to the lower jaw for chewing.

5. **Mandible**: The common name for this bone is the lower jaw. You should be able to find an upward triangular projection on the mandible that is the attachment for the chewing muscle. *Hint:* Look near the zygomatic arch.

6. **Mastoid Process**: If you feel behind the lower part of your ear, you can locate a blunt projection. Find this on the skull. Several muscles attach to the mastoid process including one that rotates and flexes the head forward.

7. **Ear Canal**: This half-funnel-shaped opening is immediately in front of the top part of the mastoid process. Pass the bristle hair into the passage and you should see it enter into the brain cavity.

8. **Sinuses**: These spongy bone areas inside the facial bones can best be seen when the bones are cross-sectioned. Open the top of the skull and look for the sinuses where the forehead bones were cut. These are the areas that become inflamed by allergies or infections.

9. **Cribriform Plate**: While you have the skull open, look at the bottom-front of the cranial cavity and find a plate with many small holes. The nerves for smelling pass through these holes in the cribriform plate.

THEN

The side view and the inside top view of the skull are shown incompletely drawn below. Draw in the missing skull structures. Your drawing doesn't have to be perfect—just make sure that it conveys the general appearance and location of the structure.

Side View	Inside Top View

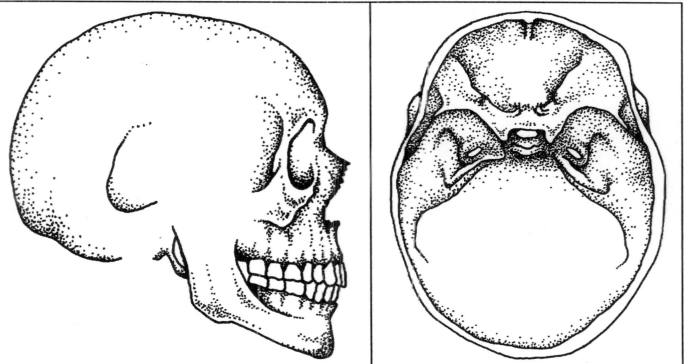

VERTEBRAL COLUMN

Twenty-six bones make up your vertebral column. It functions to support the trunk of your body and to protect the spinal cord. Each bone is called a **vertebra**. Those in the neck region are *cervical,* those in the chest area are *thoracic,* and those in the lower back are *lumbar.* The bottom two bones of the vertebral column are part of the *pelvic* structure. They are much larger because they consist of several fused vertebrae that function as a single bone.

| GO GET |  |

Go to the demonstration table marked "vertebral column."

| NOW |

Be able to identify the following structures:

1. **Vertebral Foramen:** This is the large vertical passageway through the vertebral column. It houses the spinal cord.

2. **Intervertebral Foramen:** These openings are on both sides, between the vertebrae, and serve as channels for spinal nerves to enter and leave the spinal cord.

3. **Body:** This is the part of each vertebra that actually supports the weight of your upper body.

4. **Intervertebral Disc:** These are shock-absorbing connective-tissue pads. They are located between the supporting bodies of the vertebral column.

5. **Articulations:** Look at all of the places on the vertebrae that are shaped to fit and rub smoothly against each other. Do you get an idea about how arthritis can cause back pain?

6. **Dens:** Examine the second cervical vertebra. The dens is an upward projection from the back of this vertebra that fits into the first cervical vertebra. This swivel apparatus allows you to rotate your head. Fatal whiplash occurs when the dens is driven into the top of the spinal cord. It is sometimes referred to as a broken neck, but actually the neck may not be broken.

PELVIS

The pelvis provides the sockets for the legs, which carry the weight of your body. The pelvis also includes the two fused bones of the vertebral column, and therefore supports your upper body. Because of these support demands, each pelvis half is formed by three very substantial fused bones. They are called the *ischium, pubis,* and *ilium*.

GO GET

Go to the demonstration table marked "pelvis."

NOW

Be able to identify the following structures:

1. **Ischium:**   These are the bones you sit on.

2. **Pubis:**   These are the bones that you can feel in your pubic area.

3. **Ilium:**   These are the bones that you can feel at your hips.

Label the diagram.

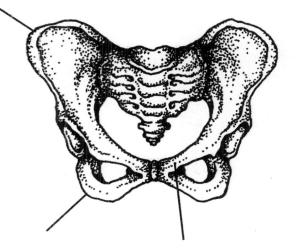

THEN

Compare the male and female pelvis.

Feature	Female	Male
Height vs. Width of Pelvis		
Pubic Angle (front underside of the pubic bones)		
Shape of Birth Canal		

LIMBS

There are 60 bones in the upper extremities (arms) and 60 bones in the lower extremities (legs).

GO GET	

Go to the articulated skeleton in the lab room, and label the skeletal diagram on the first page of the lab.

NOW

Be able to identify the following bones on the skeleton, and label the skeleton diagram on the first page of this lab.

1. **Clavicle:** the collarbone.
   Notice its attachment points.

2. **Scapula:** the shoulder blade.

3. **Humerus:** the upper arm bone.

4. **Radius:** the lower arm bone, thumb side.

5. **Ulna:** the lower arm, little-finger side.
   Notice the attachment points to the humerus.
   Which lower arm bone allows your hand to rotate?

6. **Wrist Bones:** How many are there?

7. **Femur:** the thigh bone.
   Notice its placement in the pelvis socket.

8. **Patella:** the knee bone.
   Notice its relationship to the femur and tibia.

9. **Tibia:** the shin bone of your leg.

10. **Fibula:** the smaller outside bone of your leg.

11. **Ankle Bones:** How many are there?

THE FETAL SKELETON

The fetal skeleton is in the process of formation and growth.

 GO GET

Go to the demonstration table marked "fetal skeleton."

NOW

1. Compare the relative size of the fetal skull with that of the adult skull.

2. Locate the "soft spots" (**fontanels**) in the fetal skull.
   Explain why these must exist during early human life.

3. Determine how much of the fetal skeleton is cartilage.
   What does this suggest about which comes first—*cartilage* or *bone*?

Summary Questions

1. Discuss the differences between senses and perception.

2. Discuss the relationship between two-point discrimination and the desity of touch receptors.

3. Does your body detect actual temperature or a change in temperature?

4. Name the basic parts of your ear, and tell how each part works to allow you to hear sounds.

5. How can you test whether a person has hearing loss due to middle-ear damage or nerve damage?

6. Discuss the relationship between smell and memory.

7. Discuss the evidence that taste is genetically determined.

8. Define "preferred eye."

9. How would you test whether a person normally inhibits or facilitates spinal reflexes?

SENSES AND PERCEPTION

INTRODUCTION

The human must be aware of and adjust to environmental demands. Our **senses** provide information about the physical environment. Our **perception**, developed the experiences of trial and error, evaluates sensory information, allowing us to make the appropriate adjustment to the environment. Working together, these two processes have contributed to the success of the human species.

Human senses result from specialized receptors located in various parts of the body. These *receptors* are activated by only one kind of *stimulus* (sound, touch, light, chemicals, etc.). The information from one receptor is kept separate from that sent by another sense organ. In the brain, particular areas are specialized for processing and interpreting the information pertaining to each sense.

Today's lab is designed to demonstrate some of the sensory and perceptual mechanisms of your nervous system.

ACTIVITIES

ACTIVITY #1

"TOUCH"

A string-line that loops over on the top of your head from ear to ear approximately separates the **motor area** of the brain (which controls movement) from the **touch area** of the brain (which interprets touch signals). The touch area is behind this line and the motor area is in front of it.

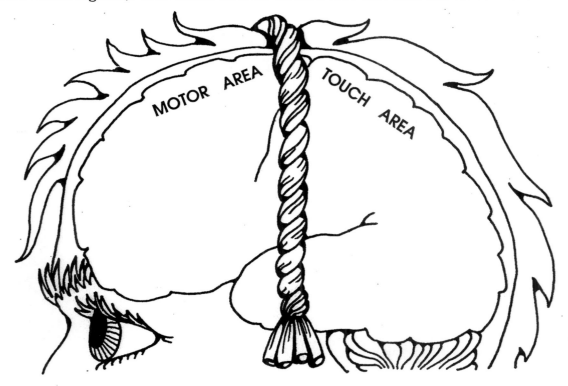

You will determine the density of touch receptors in several areas of your body. Using that information, you can test the idea that *areas of greater sensitivity contain more touch receptors*.

| GO GET |  |

1. A horse hair.

2. A metric ruler.

3. A compass.

| NOW |

1. Be sure to perform all the tests on each person in your lab group.

2. Mark off a 1 cm x 1 cm square on your fingertip, back, and another area of your body that you would like to test. You might choose an area that itches often or feels strange when touched.

264

3. Close your eyes while your lab partner lightly touches you 25 times with the horse hair. The touching should be done in a grid-like pattern that covers all of the square you have marked.

4. Each time you feel the touch of the horse hair, say so. Record the number of positive responses in the Touch Experiment Table on the next page.

5. Next, use the points of a compass to *lightly stimulate* the subject's skin in the area of the marked boxes. The compass points must be blunt and not poke through the skin. (File the points if necessary.) Start with the points close together, then increase their distance apart until the subject definitely feels *two distinct points*. Be sure that the two points are applied simultaneously each time, and retest to see if there is error due to imagination.

6. Measure the distance between the two compass points when the subject clearly perceives two points. This is called *two-point discrimination*. Record the results for each area of the body that you mapped for touch receptor density. Include the results from everyone in your lab group.

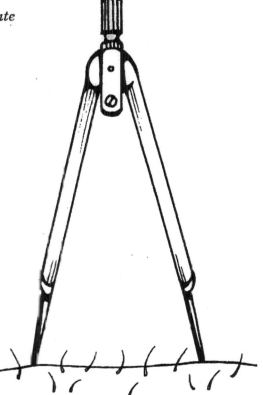

TOUCH EXPERIMENT TABLE

Part of the Body	# of Positive Responses During 25 Touches in 1 cm²	Two-Point Discrimination (in cm)
Fingertip		
Back		
(Other) _____		

? QUESTION

1. Which test area had the greatest density of touch receptors?

2. Which test area had the best two-point discrimination? Explain.

3. How is *two-point discrimination* related to *density* of touch receptors?

4. When you have an itch somewhere on your back, why does it take so much scratching before you finally find it?

ACTIVITY #2

"TEMPERATURE SENSATION"

During this Activity you will determine whether your body detects the *actual* temperature of the environment or only the *change* in environmental temperature.

GO GET

1. A large beaker of cold water (10 °C).

2. A large beaker of hot water (50 °C).

3. A large beaker of 30 °C water.

NOW

1. If the water beakers are already set up at the demonstration table, then check and adjust the temperatures using water from the hot plate or ice cubes in order to maintain the three temperature conditions listed above.

2. Place the index finger of one hand into the cold water, and the index finger of the other hand into the hot water for 15 seconds.

3. After 15 seconds, quickly place both fingers into the 30 °C water. Record the sensations.

Cold-water Finger = _____

Hot-water Finger = _____

? QUESTION

What seems to be the most important factor related to your perception of skin temperature? (circle your choice)

actual temperature   or   *change* in temperature

ACTIVITY #3

"HEARING"

The ear is divided into three parts: *outer*, *middle*, and *inner* ear. When sound waves enter the ear, the **eardrum** (between the outer and middle ear) is shaken and special small bones vibrate. These **middle ear bones** transmit the sound vibrations into the inner ear where the **auditory nerves** leading to the brain are activated. The area of the brain that is specialized for interpreting sounds is next to the ears.

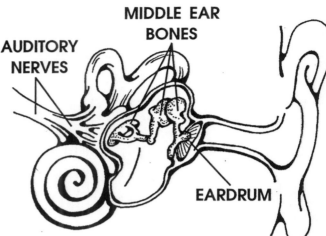

AUDITORY NERVES

MIDDLE EAR BONES

EARDRUM

GO GET

1. Some cotton for ear plugs.

2. A set of tuning forks.

3. A meter stick.

NOW

HEARING AREA

1. Do this test in a quiet room. Have the subject close one ear with cotton and close his eyes. Strike the tuning fork against the table and hold it in line with the open ear. Move the tuning fork away from the ear until the subject just loses the ability to hear it. Measure the distance. Repeat the test again to validate your first measurement. Record the hearing distance for the other ear. *Be sure to strike the tuning fork with equal force each time you do the test.*

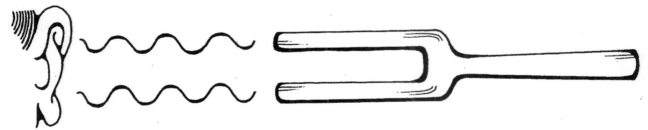

2. Repeat the test with each of the six tuning forks of different tones to determine if you have hearing loss in any of the six ranges. If one of your ears has a hearing loss at a particular tone range, then do the next test.

3. This next test should not be performed in a quiet room. Place the handle of a vibrating tuning fork on the midline of the subject's forehead.

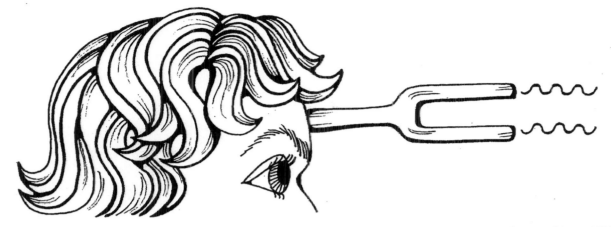

A person with normal hearing will localize the sound as if it were coming from the midline. If one ear has defective middle-ear function (ear bones), then the sound will be heard much better in the defective ear than when the tuning fork is not in contact with the forehead. If there is an affliction of the auditory nerve in the ear, then touching the tuning fork to the forehead won't improve hearing in the defective ear.

RESULTS OF HEARING TESTS

Sound Frequency (cycles per second)	Farthest Distance sound heard from Left Ear	Farthest Distance sound heard from Right Ear
128 cps fork		
256 cps fork		
512 cps fork		
1024 cps fork		
2048 cps fork		
4096 cps fork		

ACTIVITY #4

"SMELL"

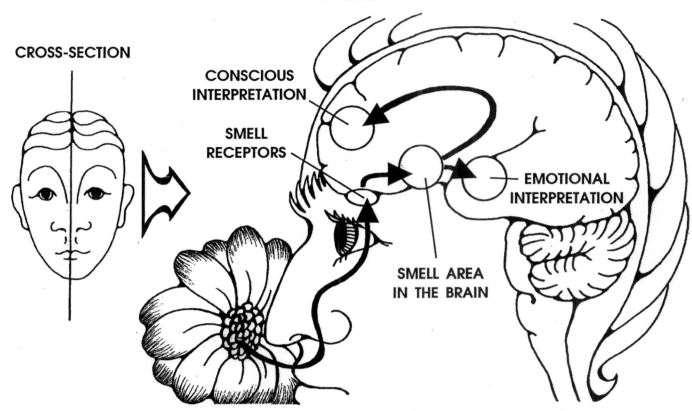

CROSS-SECTION

CONSCIOUS INTERPRETATION

SMELL RECEPTORS

EMOTIONAL INTERPRETATION

SMELL AREA IN THE BRAIN

Recent studies show that smell is much more important in human behavior than was previously thought. Some researchers suggest that the evolutionary specialization of the mammal forebrain began with the sense of smell. The exact role of smell in our lives is not understood. This sense seems to be more closely linked to emotional memories than to the conscious activities of our brains. As you experiment with the various odors in the exercise below, describe the type of *emotional reaction* you have to each.

| GO GET |  |

A smell kit.

| NOW |

1. Close your eyes. Have your lab partner pass an open odor vial about 3" under your nose for a couple of seconds. Repeat the test if necessary.

2. *First*, determine if you can smell the odor. *Second*, determine if you can correctly identify the smell. *Third*, describe any special memories associated with the smell.

3. Record the results of your test in the Odor Recognition Table.

270

ODOR RECOGNITION TABLE

Vial Number	Detects Smell	Identifies Smell	Memories Associated with the Smell
1	_____	_____	_____
2	_____	_____	_____
3	_____	_____	_____
4	_____	_____	_____
5	_____	_____	_____
6	_____	_____	_____
7	_____	_____	_____
8	_____	_____	_____
9	_____	_____	_____
10	_____	_____	_____
Totals	_____	_____	

? QUESTION

1. How many of the smells were associated with emotional memories?

2. List three examples of how specific smells might be used to sell you a product.

    a.

    b.

    c.

ACTIVITY #5

"TASTE"

The tongue has at least four different taste receptors (**salty**, **sweet**, **bitter**, and **sour**). However, the taste of many chemicals is also influenced by your interpretation of their *smell*. In this Activity, you will examine different aspects of your ability to taste.

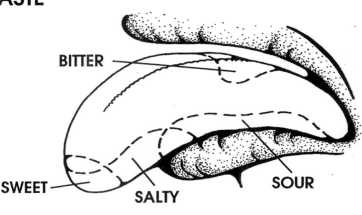

BITTER

SWEET   SALTY   SOUR

GENETICALLY DETERMINED TASTE

Your ability to taste certain chemicals is not influenced by previous eating habits, but is determined by whether or not you have inherited the *gene* controlling the taste response to that particular substance. This is an important lesson to remember when certain foods don't taste bitter to you, but other people complain about them (especially your children). They may have the gene to taste it, and you don't!

GO GET

The special taste papers for PTC and thiourea.

NOW

1. Take one taste paper, and touch it to your tongue. You will immediately know if you are a taster!

2. Do the same test with the other taste paper.

3. *Put the used taste papers in the trash.*

1. Are you a taster for thiourea?

2. Are you a taster for PTC?

3. If you are a non-taster and you want to be a high-class chef, what might you do to compensate for this genetic limitation?

SUGAR TASTE THRESHOLD

This experiment should give you insight about why some people prefer more sugar in their foods.

NOW

1. Go to the demonstration table and determine your sugar taste threshold.

2. Dip a strip of tasting paper into each solution, and record whether or not you can detect a sweet taste. *Discard the used taste papers.*

3. After testing all of the solutions, go to the front desk for the key to the sugar concentration of each solution.

4. Record your results in the summary chart on the front chalkboard.

? QUESTION

1. What was the lowest threshold for tasting sugar?

2. What was the highest threshold?

3. Do the people with a high taste threshold also like to add more sugar to their food? (Perhaps you could determine this by asking your classmates how much sugar they add to their coffee.) Your lab may have a setup for making a "food coloring" tongue print to count the number of taste buds in different people.

ACTIVITY #6

"VISION"

Human beings are primarily visual animals. This is the dominant sense you use to relate to the environment. Furthermore, most human behavior has been shown to be strongly influenced by visual perception. There is a lot of scientific literature on visual perception and we encourage you to investigate this information when you have time to do so. What you don't know can be used against you!

PREFERRED EYE

This Activity is designed to reveal which one of your eyes is used for certain visual functions. Your *preferred eye* is the one your brain chooses to use when both eyes can see the same object.

NOW

1. Pick an object that's about 30 feet away. Make a circle with your thumb and first finger of both hands.

2. Straighten and raise your arms from your waist to a position where the circle surrounds the object. Keep your head and feet positioned straight ahead.

3. Without further movement, close one eye. Then open the closed eye, and close your other eye.

4. Which eye has the *same view* as the view with *both eyes* open? This is your preferred eye. *Hint:* When you close your preferred eye, the distant object will move out of the circle formed by your hands.

? QUESTION

1. Which eye is your preferred eye?

2. If you are left-eyed, what problem will you have in shooting a rifle?

3. Why should you use your preferred eye when looking through a monocular microscope?

EYE WITH BEST VISION

Use the classroom eye chart to determine which of your eyes has the best vision (without glasses).

? QUESTION

1. Which of your eyes has the best vision?

2. Talk with other lab students, and discover whether the eye with the best vision is always the same one as the preferred eye. Results: _____

EYE WITH BEST DEPTH PERCEPTION

There are fairly simple ways of determining which of your eyes has the best depth perception. If this equipment is available, then determine the depth perception for each of your eyes. If this equipment is unavailable, then refer to the information chart to answer the questions below.

INFORMATION

Vision Tests	13 Left-Handed People	11 Right-Handed People
Eye with Best Depth Perception	9 left eye 3 right eye 1 same in both eyes	2 left eye 8 right eye 1 same in both eyes
Preferred Eye	7 left eye 6 right eye	5 left eye 6 right eye
Eye with Best Vision	1 left eye 1 right eye 11 same in both eyes	2 left eye 1 right eye 8 same in both eyes

? QUESTION

The eye with best *depth perception* is most closely associated with . . . (circle your choice)

Preferred Eye   or   Preferred Hand   or   Eye with Best Vision

ACTIVITY #7

"REFLEXES"

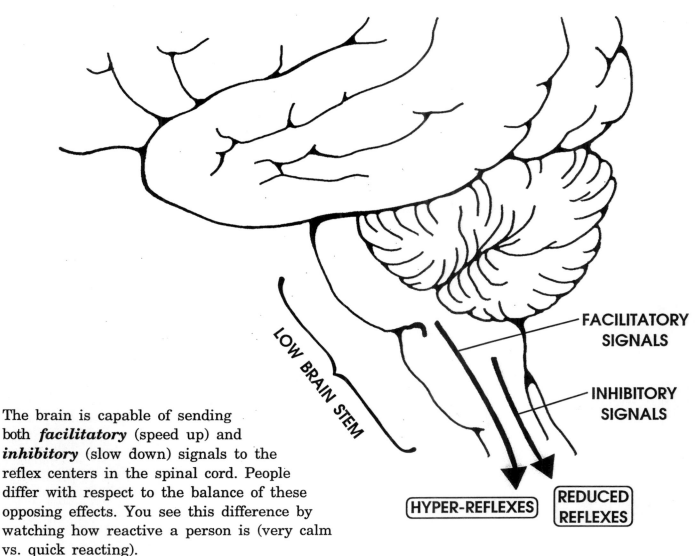

FACILITATORY SIGNALS

INHIBITORY SIGNALS

LOW BRAIN STEM

HYPER-REFLEXES

REDUCED REFLEXES

The brain is capable of sending both *facilitatory* (speed up) and *inhibitory* (slow down) signals to the reflex centers in the spinal cord. People differ with respect to the balance of these opposing effects. You see this difference by watching how reactive a person is (very calm vs. quick reacting).

A reading reflex test can be used to determine which reflex type you are. *Concentrating on reading should reduce the effect your brain normally has on your reflex centers.* If reading reduces your reflex response, then normally your brain must be stimulating reflexes (you are a quick-reacting person). If reading increases your reflex response, then your brain normally inhibits reflexes (calm reacting). Your brain's reflex emphasis can change. No person is 100% one type or the other all of the time.

GO GET

1. A patellar hammer.

2. A meter stick.

1. Before beginning the test, ask your lab partner to evaluate whether you are the calm or quick-reacting type.

    Lab Partner's Opinion: _____

    Your Opinion: _____

2. Sit on a table so that your legs hang freely over the edge. Have your lab partner hit the patellar ligament (just below the knee) with the reflex hammer. *Don't hit too hard.* This may take some practice. Measure the amount of leg movement several times in order to get an average estimate of the reflex intensity.

    Normal Reflex: _____

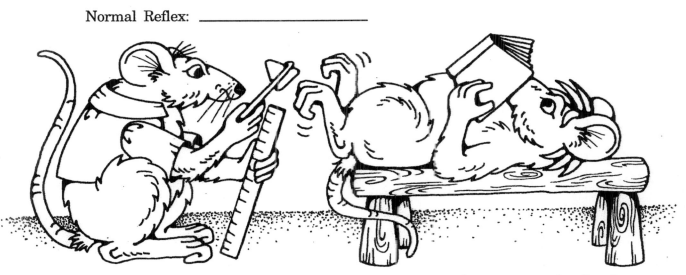

3. Next, read from a textbook while your lab partner measures the amount of reflex leg movement. Is the reflex more intense or less intense during the reading conditions?

? QUESTION

1. Under normal circumstances, does your brain activate or inhibit spinal reflexes?

2. So, which reflex type are you?

3. Does this agree with how you evaluated yourself before the test?

4. Compare your conclusions with those of other students in the class. What did you discover?

Summary Questions

1. Discuss the difference between intuition and empirical thought.

2. What is meant by the term "counter-intuitive?"

3. In a low-risk situation, how do most people act?

4. In a high-risk situation, how do most people act?

5. What two things happen when there is ambiguous or conflicting information surrounding a particular decision-point?

6. How good is our intuition at understanding things that are in a different scale from those in our personal lives? Why?

7. What is the "Rule of 70?"

8. What is the economic definition of "cost?" Explain what is meant by that definition.

9. Discuss "knowing that you are right."

ARE HUMANS DOOMED TO INTUITION?

INTRODUCTION

People often make the comment, "Problems in the modern world are out of control." Furthermore, they say, "If it weren't for science telling us what to do, everything would be better."

Those who understand the methods of science find these comments maddening. The everyday decisions controlling government, social institutions, and individual freedom are not made using the empirical method of science. Instead, society usually searches for solutions and reaches decisions by using human intuition.

This week's lab will focus on *intuitive* and *counter-intuitive* processes, and the role that science could play in our society's future decisions if we began to use it.

ACTIVITIES

ACTIVITY #1

"QUESTIONNAIRE"

It is important for you to answer all of the following questions by yourself (*with no discussion at all*) before starting the other Activities in this lab. The answers to these questions will give you valuable insight into your own thinking processes, and will be used as part of the class data for group comparisons during the lab.

QUESTIONS

1. Define what you mean by the word "intuition."

2. What is the source of intuition?

3. Do you use intuition to make decisions?

4. If so, in what kinds of decisions?

5. Can intuition be wrong?

6. If so, when is it wrong?

7. Don't analyze. Make a quick decision after reading the following information. There are two buttons: **A** and **B**. You must push one of them. (circle your choice)

    Button A—If you push this button, you will be given $3000.

    Button B—If you push this button, you have an 80% chance of getting $4000, and a 20% chance of getting nothing.

8. Don't analyze. Make a quick decision after reading the following information. There are two buttons: **C** and **D**. You must push one of them. (circle your choice)

    Button C—If you push this button, you lose $3000.

    Button D—If you push this button, you have an 80% chance of losing $4000, and a 20% chance of losing nothing.

9. How many people do you usually date before forming a serious relationship?

10. How many different college majors are you considering as definite possibilities?

11. In your opinion, do most people stay in bad relationships longer than they should?

12. In your opinion, do most people stay in unsatisfying jobs longer than they should?

13. Your brain makes some decisions without your conscious awareness. What could you do to discover this?

14. If you observed that all humans make the same mistake, what would you conclude about human behavior?

15. How many people live in the USA?

16. Using the population of California as a base, how many *additional people* are added to the world each year? (That is, how many "Californias" are added to the planet each year?)

17. Using the population of Los Angeles as a base, how many *additional people* are added to the Western Hemisphere each year? (That is, how many "LAs" are added to our half of the planet each year?)

18. You have decided to buy an $18,000 new car, and have been offered three methods of payment by the dealer. Quickly estimate the total cost of each payment method.

   Payment by cash = _____

   5-year 10%-interest car loan = _____

   Add the car payment to your
   existing 8% home-equity loan = _____

19. How many months of work will it take you to pay for a new $18,000 car if the method of payment is cash?

20. Compared to bicycle travel, how much time in a year is saved when traveling by car?

21. When you make decisions, how important is the intuitive feeling of "knowing that you are right"?

ACTIVITY #2

"INTUITION AND COUNTER-INTUITION"

Several concepts must be understood before we can continue our investigation. First, we need some clear definitions. The dictionary describes *intuition* as (1) the act or faculty of knowing without the use of rational process; (2) a capacity for guessing accurately; and (3) a sense of something not deducible.

These definitions suggest that intuition does not depend on analysis of a situation, and that intuition is usually correct. In addition, the definitions imply that certain truths may be discovered only through intuition and *cannot* be known by using logical processes.

The primary process in science is called empirical thinking. *Empirical* is defined as (1) relying upon or derived from observation or experiment; and (2) not guided by theory or intuition, but by verifiable measurements or observations.

When science tests intuitive ideas, there are two possible results:

1. The intuitive idea agrees with the empirical evidence (science).

2. The intuitive idea does not agree with the empirical evidence (science). The processes that science describes differently from intuition are called *counter-intuitive*.

One example of a counter-intuitive process is the *rotation of the Earth*. Our intuition tells us that the sun revolves around us. We see it rise in the east and set in the west. However, science has proven that our intuition is wrong. The planet spins on its axis from west to east, and one complete rotation takes 24 hours.

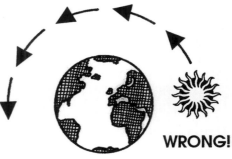

WRONG!

Dreaming is another example of a counter-intuitive process. Dreams usually involve strange places and situations, yet our friends tell us that we haven't physically moved from our beds at night. Our intuition leads us to conclude that there is a spirit world that we go to when we are sleeping. However, science has proven that dreams result from various spontaneous activities in the brain that can be stimulated or suppressed by machines or drugs. Dreams apparently have nothing to do with a spirit world.

There are *two* useful ways of describing the consequences you may experience when some aspect of the world operates counter to your intuition.

COUNTER-INTUITIVE PROCESS

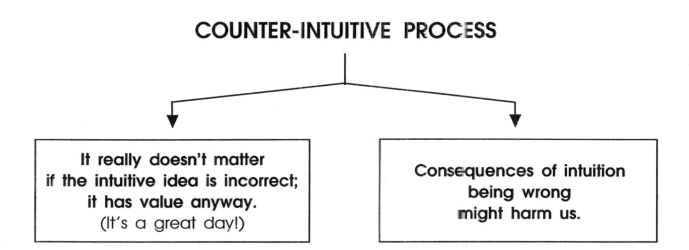

| It really doesn't matter if the intuitive idea is incorrect; it has value anyway. (It's a great day!) | Consequences of intuition being wrong might harm us. |

While you are doing the following Activities, consider what serious consequences might occur when intuition is wrong. Also consider if there might be a scientific explanation for why humans don't recognize and solve certain world problems. Finally, consider what would happen if we used scientific thinking to help us when our intuition fails.

? QUESTION

1. Define intuition.

2. Define empirical.

3. Define counter-intuitive.

IMPORTANT MESSAGE

Remember: The concepts in this lab require your complete understanding of the definitions for intuitive, empirical, and counter-intuitive thought. Check with your instructor now if you are confused.

ACTIVITY #3

"DECISIONS"

Can we agree that the ability to make a decision is an important human act? Can we agree that your relationships, your choice of a college major, and how you spend your money are important decisions? If you answered "yes," then let's look at the discoveries made by two scientists, Daniel Kahneman and Amos Tversky, who have studied human decision-making.

CLASS DATA

Refer to the "Button Problems" in the Questionnaire, Activity #1. Record your answers to questions #7 and #8 on the chalkboard. Write the class totals below.

Total # of Students in Your Class = _____

Choosing Button A = _____

Choosing Button B = _____

Choosing Button C = _____

Choosing Button D = _____

Kahneman and Tversky found that most people choose Button A and Button D. Do these findings agree with your class results? _____

Do the findings agree with your personal choices? _____

A *low-risk situation* is one in which you are not going to lose something that you already have.

A *high-risk situation* is one in which you are going to lose something whatever choice you make.

? QUESTION

1. Which situation is low risk? (circle your choice)

     A and B   or   C and D

2. Which Button choice in the low-risk situation is "playing it safe"?

3. Which Button choice in the low-risk situation is "taking a chance"?

4. So, how do most humans act when they are in a low-risk situation? (circle your choice)

     They play it safe.   or   They take a chance.

5. Which Button choice in the high-risk situation is "playing it safe"?

6. Which Button choice in the high-risk situation is "taking a chance"?

7. So, how do humans usually act when they are in a high-risk situation? (circle your choice)

     They play it safe.   or   They take a chance.

LET'S EXAMINE THE FACTS

The button decision problems were designed to offer you a choice, and determine whether you have an intuitive bias. The results reveal that intuition directs us to choose A and D, both of which are *losing* strategies in life. The *winning* choices are actually B and C.

In a low-risk situation there is no real personal loss to you if you took a chance. You were being given money. The odds favored more money if you chose Button B. Your choice of Button A means that you stayed with the safe bet. It's just like staying with one major in college or dating only one person. *You are limiting your options for better returns.* After all, who would insist that there is only one deserving person in the world with whom you could have a successful relationship? Likewise, our economy proves that there are many successful choices for college majors. But, in money, love, and vocation we usually find ourselves playing it safe in a world of opportunity.

By the way, if you happened to choose Button B and you're feeling pretty smug right now, then forget it. Research has shown that when the problem is restated in terms other than money, you will make the same mistake as everyone else (Button A). For example: You are hungry. Button B offers you, free of charge, a seven-course dinner at a posh restaurant with a 20% chance that the restaurant is closed. Button A offers you, free of charge, a meal at your favorite fast-food emporium that is open 24 hours a day, seven days of the week.

A more serious problem is revealed by the selection of Button D. Both Buttons C and D involve loss. However, the intuitive bias towards Button D means that you are willing to expose yourself to *even more risk* in an effort to avoid any pain. Choosing Button D compounds the risk. Choice C, while still a loss, represents the best strategy—*take the smallest loss and move on.* Moving on to a new relationship has a greater chance of success than waiting for a bad relationship to improve. Getting a new job has more possibility of encouraging the development of your talents than staying at an unsatisfying job.

NOW

1. Divide the class into discussion groups. Assign one person the responsibility of coordinating the discussion. (This person is not supposed to come up with all of the ideas—just keep the rest of the group on task.) Assign another person to write down the group's answers.

2. For each of the three topics, determine how your using strategy A or D to solve the problem could actually make the situation worse.

   A = *Playing it safe when there is opportunity to take a chance without loss.*

   D = *Taking even more risk in an attempt to avoid loss.*

3. For each of the three topics, determine how your using strategy B or C to solve the problem might improve the situation.

   B = *Taking a chance for a new opportunity when there is no real loss at stake.*

   C = *Taking the loss and moving on to new opportunities.*

4. Work on the three topics for 10 minutes each. There will be a class discussion following your group discussions.

GROUP REPORT

TOPIC #1   HUMAN OVERPOPULATION

1. Using **Strategy A**, what political policy might be adopted towards human population growth well before there is a serious overpopulation problem?

   What are possible *negative* consequences of using this strategy?

2. What is an example of a **Strategy D** political policy towards population growth once overpopulation begins to cause serious problems?

   What are possible *negative* consequences of using this strategy?

3. What is an example of a **Strategy B** political policy towards population growth today before there is a serious overpopulation problem?

   What are possible *benefits* of using this strategy?

4. What is an example of a **Strategy C** political policy towards population growth once overpopulation begins to cause serious problems?

   What are possible *benefits* of using this strategy?

TOPIC #2   UNEMPLOYMENT

Try another example using a separate sheet of paper. What are examples of the four strategies an unemployed person could have used prior to a factory shutdown, or while they are on unemployment benefits?

TOPIC #3   PERSONAL RELATIONSHIP

Finally, try personal examples using the four strategies to find a good relationship, or while in a bad relationship.

RANDOM DECISIONS DURING PROBLEM-SOLVING

Many aspects of decision-making and problem-solving have been investigated by science. These problem-solving situations have been tested and retested by researchers, and the conclusions are similar whether a single person, a family, a company, or a governmental agency is studied. Scientists have discovered the influence of *random thinking processes* during problem solving.

The first step the research team took was to methodically interview a test group. This was to determine what different *decision-making criteria* the group used when solving a particular problem. The City Council flow chart below is a typical example of the decision-making process.

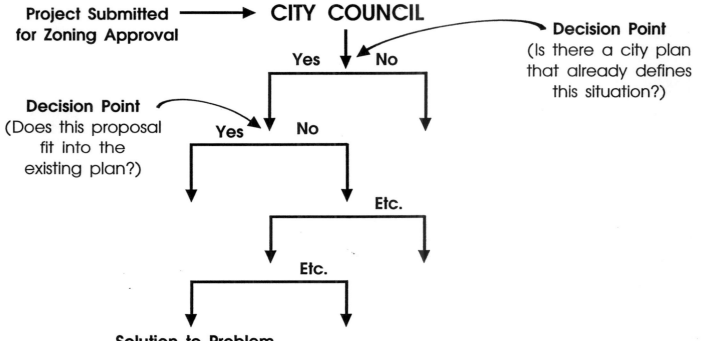

Once all of the test group's decision-making criteria had been determined, the researchers began submitting problems to the group to see if they actually follow their own rules.

RESULTS

1. All groups developed solutions that included *violations* of their original decision-making criteria.

2. When the decision-points were carefully monitored, the researchers discovered that the test groups were completely *unaware* that they had "violated" their own rules.

3. The violated decision-points were retested. Researchers determined that there was no pattern. The test groups *randomly chose* between "yes" and "no," and they were unaware that their choices were random.

4. Further investigation revealed that the test groups made random choices whenever there was *ambiguous* or *conflicting information* surrounding a particular decision-point.

CONCLUSION

There is something inherent in the human thinking process that automatically creates random decisions in the brain. And, whenever there is ambiguous or conflicting information, the number of random choices increases, and *the brain doesn't realize it!*

NOW

1. Divide the class into discussion groups. Answer the questions below keeping in mind what researchers learned about decision-making. Assign one person the responsibility of coordinating the discussion. Assign another person to write down the group's answers to each of the eight questions. Be very specific about answering question #7.

2. Your group has 20 minutes to discuss all the questions. Then the class will have a general discussion for 10 minutes.

? QUESTION

1. When people have strong disagreements in personal relationships or involving group decisions, is it likely that there will be ambiguous or conflicting information at the decision-points?

2. What is going to happen in both people's minds as a result of the answer to question #1?

3. Will they be aware of this process during the discussion?

4. How are they likely to treat each other during the problem-solving?

5. Based on what you have learned in this exercise, would you expect humans to be able to work out their personal relationship problems by using the usual human thinking processes?

6. Would you say that your society has the same problems as you do?

7. What is your group's solution to the dilemma illustrated in questions #1–6?

8. Use the same approach as you used to answer question #7 to project how our society might solve the problem of overpopulation.

ACTIVITY #4

"PROBLEMS OF SCALE"

We have a great deal of difficulty understanding processes that are many times larger or smaller in scale than those in our personal lives. You were asked three questions about population (questions #15, #16, and #17) in the questionnaire at the beginning of this lab.

The answers to the three questions might surprise you.

#15: There are 265 million people in the United States.

#16: Three times the population of California are added to the world each year.

#17: Three times the population of Los Angeles are added to the Western Hemisphere each year.

Most people miss these estimates unless they have recently seen them in a magazine article. Apparently, we do not comprehend population numbers because they are not in the scale of our normal, everyday experience. Therefore, the process of analyzing population growth is *counter-intuitive*. One way of revealing this fact is to examine our intuitive inadequacy in understanding the **rate of growth** and **doubling times**.

There is a rule and a diagram that can help us to go beyond our intuition when we try to solve a problem involving a rate of growth (business planning, investments, debt, population, etc.).

RULE OF 70

If you divide the number 70 by the growth-rate percent of a process, then you will have estimated the number of years required for that process to double. For example: If the national debt is increasing at a rate of 7% per year, then the debt will double in $70 \div 7 = 10$ years.

THE DOUBLING RECTANGLE

There is a visual aid to help you understand the doubling process. It is called the **Doubling Rectangle**, and it looks like this:

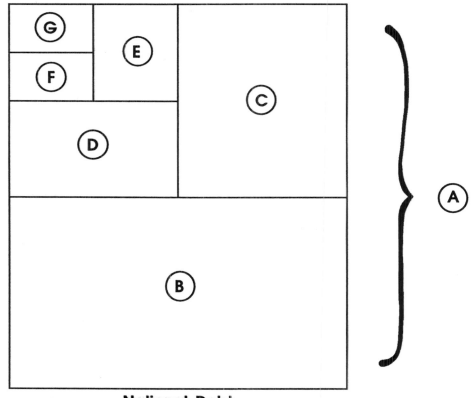

National Debt

Use the Doubling Rectangle as a visual aid to understanding the doubling process of the National Debt. The easiest application of the *Doubling Rectangle* is when the growth rate has been constant over a long period of time. Let's assume that the growth in national debt has been constant and that each rectangle represents 10 years of time. Label rectangle Ⓓ as the 1980s decade. That means that Ⓒ, Ⓑ, and Ⓐ are the doubled debts for the 1990s, from 2000–2010, and from 2010–2020. Going back in time, Ⓔ becomes the decade of the 1970s, Ⓕ the 1960s, and Ⓖ represents all debt in the United States before 1960. Label all of the rectangles in the diagram with these decades. *You will be using this Doubling Rectangle to compare with the one that you do next.*

1. Try a problem of scale using the tools you have just learned. The GDP (Gross Domestic Product) has been growing at an average of 3% per year since 1900. Your job is to calculate the doubling time, and then label the *Doubling Rectangle* below.

2. Start with rectangle Ⓒ, and label it **1970** ⟶ **?**. The **?** will be known after you calculate the GDP doubling-time using the *Rule of 70*. GDP doubles in _____ years.

3. Rectangle Ⓑ will start with the year that rectangle Ⓒ ended with. Label the doubling time for Ⓑ.

4. Finish the diagram after doing the calculations. Label Ⓐ, Ⓓ, Ⓕ and Ⓔ. Rectangle Ⓖ represents the GDP before 1900.

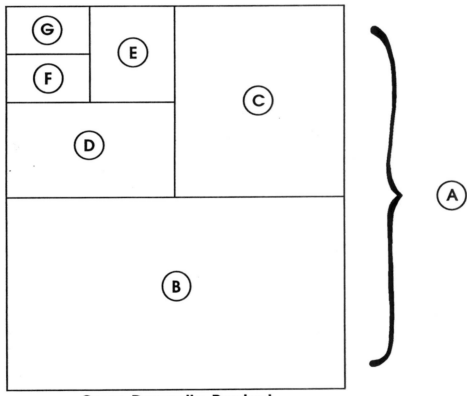

Gross Domestic Product

Check with your instructor to make sure that your answers for doubling times are correct before going on to the next questions.

1. Which process has the quickest doubling time? (circle your choice)

   GDP   or   National Debt

2. You should have calculated that the GDP rectangle (B) starts in 1993. During the time when the GDP is doubling in (B), how many doublings occur in the National Debt? (Refer to your first doubling rectangle.)

3. The period between 1940 and 1960 was the time of fastest growth in personal income and lifestyle for Americans. Economists tell us that the GDP grew at a rate of 4.4%, which is faster than the average for the century. What is the GDP doubling-time based on a 4.4% growth rate?

   Was the growth in the GDP closer to the growth of debt during the 1940–1960 period than it has been since then?

4. We have been increasing our use of oil at a rate of 7% per year since the turn of the century. What's the doubling time?

   Pick any decade (such as 1980–1990). The amount of oil consumed in that decade is how much compared to all of the oil used before that decade?

5. The world population is growing at about 2% per year. What's the doubling time?

6. The doubling time in food production for both the USA and Mexico has been about 35 years. Mexico's population doubling-time has been about 25 years and the USA's has been about 50 years. What has Mexico had to face during the past three decades that the USA has not had to contend with?

7. Since 1990 the average wage has been increasing at 1.5% per year. (This does not consider people who lost higher-paying jobs and were rehired at lower-paying ones.) The rate of inflation has been increasing at about 3% per year. At these rates, how long will it be before your real spendable income becomes one-half what it is today?

Doubling times cannot be accurately comprehended by using human intuition. As you saw in the previous examples, counter-intuitive problems are complex, and we haven't considered that (1) doubling times are not constant, and (2) there are many other variables besides the GDP and debt that need to be considered at the same time.

IMPORTANT MESSAGE

Perhaps you have heard of the computer game called *SimCity*, based on city planning. The fun of this game is that the dynamics of the game are based on rates of change (doubling times). While playing the game, we discover that our intuition is continually being "fooled."

What is fun in a game can be a catastrophe in real life!

ACTIVITY #5

"PROBLEMS OF COST"

Modern economic theory states that the free-market capitalistic system operates efficiently because of rational self-interest. ***Rational self-interest*** means that we make decisions that are in our best interest. Economists tell us that rational self-interest creates a natural buying and selling market that balances supply and demand so that stable prices result. One test of these assumptions is to ask the question: "Can our intuition balance the costs and benefits of our purchasing decisions?"

Economics textbooks define ***cost*** as ***the value of opportunities forgone in making choices***. One simple test of our ability to determine monetary costs is found in question #18 of the questionnaire. Let's check your accuracy. The cost of a car when paying in cash is $18,000. The cost of using the 5-year 10%-loan is $24,000. The cost spread over a 30-year 8% home-equity loan is $54,000.

? QUESTION

1. How close were your questionnaire estimates to the actual costs?

2. Was your intuition adequate to the task?

NOW

Convert the cost of a new car into the number of months you must spend working to pay for it.

1. What wage do you expect to be making? ($ per hour) = _____.

2. Assuming 160 hours of work per month, your Gross Monthly Wage = _____.

3. Subtract your taxes (about 35% of income). Corrected Monthly Wage = _____.

4. Subtract your estimated rent and food. Corrected Monthly Wage = _____.

5. Subtract routine miscellaneous expenses. Corrected Monthly Wage = _____.
   This is your *Spendable Income*.

6. Divide the cost for each of the three methods of car purchase by your Spendable Income, and put your answers in the table on the next page.

297

Method of Payment	Months of Work to Pay for Car
$18,000 ÷ Spendable Income	_____
$24,000 ÷ Spendable Income	_____
$54,000 ÷ Spendable Income	_____

? QUESTION

1. Do the calculated values for the amount of work required to pay for your car surprise you?

2. If so, what does this mean about your intuitive ability to understand costs?

Remember: The time spent working for the car is a measure of cost (opportunities forgone by making the decision to buy a new car).

TRAVELING BY CAR OR BY BICYCLE

Could driving a car cost you as much time in your life as if you traveled by bicycle? Set your intuition aside and give the environmentalists a chance to convince you.

INFORMATION

1. The average travel by car in the USA is 12,000 miles each year. The environmentalists are correct when they say that many European cities are designed for bicycle travel. Their cities are efficiently planned so that an average person needs to travel only 6,000 miles per year to do everything we do in 12,000 miles.

   If you drive a combination of city and open-road, then divide 12,000 by 25 mph (your average speed), and put this value into the Travel Comparison Table (travel time by car). If you drive only in the city, then add another 320 hours of time because your average speed drops to 15 mph in a big city.

   Traveling by bicycle averages 10 mph. Divide this into the 6,000 miles that you would need to commute in a well-designed city. Record that number in the Travel Comparison Table (travel time by bicycle).

2. The average expense of operating a car is $3,900 per year (car cost, repair, insurance, gas, etc.). In addition, government estimates reveal that about $1,000 of tax money is spent subsidizing the automobile each year for each driver (highway construction and upkeep, parking, environmental expenses, etc.).

   Divide the total yearly expense for operating a car ($4,900) by the hourly wage you chose for the car cost problem on one of the previous pages. Record your answer in the Travel Comparison Table (expenses).

3. The average minimum expense of operating a bicycle is estimated to be $500 per year (bike cost, repair, parking, road upkeep, and injury insurance). Divide the total yearly expense for operating a bicycle ($500) by your hourly wage. Record your answer in the Travel Comparison Table (expenses).

4. Calculate the totals for each method of travel.

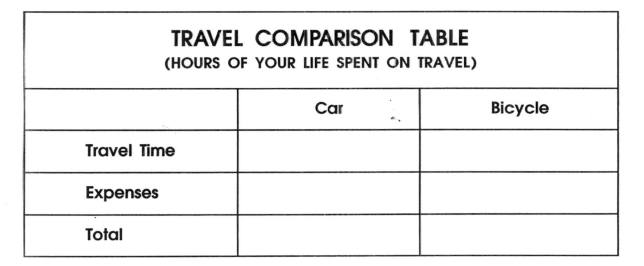

TRAVEL COMPARISON TABLE (HOURS OF YOUR LIFE SPENT ON TRAVEL)		
	Car	Bicycle
Travel Time		
Expenses		
Total		

Are you surprised that travel by bicycle compares so well to travel by car?

Bicycle travel is *counter-intuitive* for Americans. This comparison does not prove that car travel is inferior, nor does it prove that we should all travel by bicycle. However, our study does demonstrate that by using our intuition, the estimate of actual cost of travel will be incorrect (another opportunity forgone).

ACTIVITY #6

"KNOWING THAT YOU ARE RIGHT"

The final Activity of this lab examines the intuitive feeling of *knowing that you are right* in a particular situation. Your group will solve a very common problem in city redevelopment. ***An area of the city does not have enough jobs, and there are not enough cheap houses for the lower-income people who live there.***

NOW

Your group is to quickly write down your solutions for city redevelopment, and predict the positive results of your plan. After you have done this, read the information below.

Don't read the Empirical Findings until after your group has devised a solution to redevelopment. If you cheat, you won't get as much value from testing your intuition (an opportunity foregone).

EMPIRICAL FINDINGS

The most common solution to the city redevelopment question is to build more cheap housing. This solution always fails. More lower-income people are attracted to the area, and *the local unemployment actually increases*. Another common solution is to build factories in the area. This approach works only when building the factories also reduces the existing housing and number of residents at the same time. An unexpected finding is that destroying low-income housing will increase the percentage of employed people remaining in the area. These results have been known for several decades, yet the same failed approaches are continually recommended by politicians and planners.

Review your group's solution again. Do you still think that your plan would work if only it was given a little time and a fair chance?

ANSWER

Most people answer "yes" to this question. It is almost universal that when people are presented with the empirical evidence that disproves their intuitive idea, they still "feel" that they are correct.

The intuitive feeling of *knowing that you are right* apparently does not change even when the facts oppose you.

This is the final challenge to humanity as it considers those aspects of the modern world that are counter-intuitive. To be successful as a species, we may have to ignore our intuitive feelings when solving certain problems—and that could be very difficult.

How will you play the game of life: Take your losses and move on, or risk even more? There is a lot to be said for understanding the real world choices behind Buttons A, B, C, and D. Ultimately, your future will be determined by the buttons you choose.

1. How does an embryo differ from a fetus?

2. What is meant by the phrase "ontogeny recapitulates phylogeny?"

3. During which days of the menstrual cycle would you expect:
   immature egg
   ovulation
   corpus luteum
   shedding of endometrium

4. Discuss the differences between the developmental stages of male and female reproductive systems (eggs and sperm) in the fetus and at puberty.

5. If the female eggs are damaged by environmental factors, can they return to normal over time?

6. If the male sperm are damaged by environmental factors, can they return to normal over time?

7. What tubes are surgically cut in males and in females as a means of birth-control?

8. What is the function of the placenta?

9. Embryonic development can go wrong. The general rule is, the _____ the disruptive event, the _____ the consequences.

10. List the effects on embryonic development caused by alcohol, by cigarettes, and by poor nutrition.

EMBRYOLOGY

INTRODUCTION

Embryology is the study of early developmental phases in plants and animals. An **Embryo** is any stage after the egg becomes fertilized, but before the developing organism looks like the adult form (about 8 weeks for humans). In advanced vertebrates, any stage after the embryo and before birth is called a *fetus*.

A century ago, there were two contrasting theories about development. One view (*preformation*) held that a miniature version of the adult existed in the egg and grew into the adult form. This theory is incorrect. The second theory (*epigenesis*) stated that new structures and body systems are continually being created during embryonic development. This description is correct. All vertebrates begin as a single cell, but develop amazing degrees of complexity by the time they are born.

Scientific research has revealed two very important embryological concepts. First, the study of embryos shows how simple life-forms could have evolved into complex life-forms. The developing traits provide biologists with the opportunity to observe how genetic mechanisms operate to create an organism. The second concept is that there are developmental processes common to all vertebrates. This means that studying the embryos of simpler organisms can help us to understand development in human embryos.

In today's lab, you will learn the general terminology and descriptions of embryonic stages, and you will compare the embryos of various organisms at similar stages found in human development.

ACTIVITIES

ACTIVITY #1

"ONTOGENY RECAPITULATES PHYLOGENY"

The title of this Activity has been a recurring TV quiz-show question. The quote is from Ernst Haeckel, a German scientist at the turn of the 20TH century. He thought that embryonic development replayed the entire evolutionary history of a species. In other words, watching the development of a frog would be like seeing a movie of the evolution of vertebrates leading to the frog. Since Haeckel's time, science has discovered that both evolution and development are far more complicated than his original idea suggested.

The human embryo does proceed through various levels of complexity similar to those of earlier vertebrates. There is certainly as much change in the human embryo from conception to birth as there is change in the fossil record from one-celled organisms to complex land vertebrates. The situation faced by the human embryo is similar to the challenges presented to vertebrate groups that evolved from a water environment to dry land conditions. But, we do not first become a fish, then a frog, followed by a reptile, and finally a mammal. The developmental "movie" is much more blurred than that.

NOW

1. Read the brief descriptions of human embryonic stages on the next page, and then refer to the "Ontogeny Recapitulates Phylogeny" chart on the following page.

2. Then, complete the chart by writing the name of the appropriate human embryonic stage that compares in complexity to organisms in evolutionary history.

HUMAN EMBRYONIC STAGES

Name	Description
Zygote	A diploid cell formed by the union of egg and sperm.
Morula (3–4 days)	A small solid ball of identical cells.
Blastocyst (1 week)	Cells begin differentiating into tissues.
25 Day Embryo	The heart begins to beat, but there are no organ systems yet (i.e., circulatory systems, etc.).
4 Week	The *notochord* has formed. This is the very beginning of a skeletal system that will develop in later weeks.
4–5 Week Embryo	Organ systems have developed including a primitive type of kidney called the *mesonephros*. This kidney will soon degenerate and be replaced by a new one.
5–6 Week Embryo	Organ systems are developing very rapidly. A new type of kidney appears called the *metanephros*. This will eventually grow into the adult human kidney.
6–8 Week Embryo	There is a rapid development of a brain lobe called the telencephalon which grows into the *cerebrum* in later weeks.
8 Week Fetus	This stage is now called the *fetus* because it looks like the adult form.

| **ONTOGENY** | **RECAPITULATES** | **PHYLOGENY** |
| (Developmental Stages) | (Repeat) | (Evolutionary History) |

_____

Pig and Primate Embryo
(developed cerebrum of brain)

↑

_____

Chick Embryo
(has dry land kidney called *metanephros*)

↑

_____

Frog Embryo
(has aquatic kidney called *mesonephros*)

↑

_____

Early Vertebrates
(beginning of internal skeleton)

↑

_____

Flatworms and Roundworms
(definite organs present)

↑

_____

Jellyfish
(cells organize into tissues)

↑

_____

Colonial Protozoa
(cluster of identical cells)

↑

_____

Eukaryotic Cells
(unicellular with organelles)

↑

(no comparable stage
in human development)

Prokaryotic Cells
(no cell organelles)

1. Define *embryo*.

2. Define *fetus*.

3. What week of development marks the beginning of the fetus stage in humans?

4. What is meant by "ontogeny recapitulates phylogeny"?

5. Is it true that the human embryo first looks like a fish, a frog, and then a reptile before reaching the fetal stage? Explain your answer.

ACTIVITY #2

"EGGS, SPERM, AND FERTILIZATION"

EGGS

By the fifth week, a human female embryo already has pre-egg cells multiplying in her ovaries. These pre-egg cells begin meiosis and develop into a cell stage called the *oocyte*. Each oocyte is surrounded by a small cluster of cells called the **follicle**. There are about a million oocytes in the ovary of a female baby at birth. These immature eggs are stopped at Prophase I of meiosis until she reaches puberty. This is a significant vulnerability because the chromosomes of oocytes are exposed to many environmental chemicals during childhood. Chromosomes can be damaged by these chemicals.

The oocytes mature into eggs beginning at puberty. Usually one egg is released (called **ovulation**) each lunar month. This monthly process, called the **menstrual cycle**, is initiated by follicle stimulating hormone (**FSH**), which is produced by the pituitary gland (one of the glands of the endocrine system).

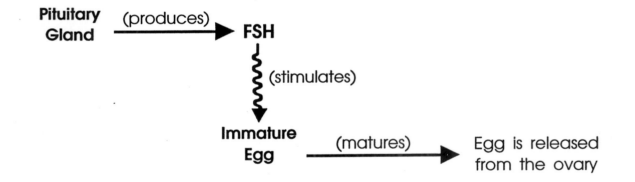

As the egg matures, the other cells surrounding it divide and grow. Most of this growing cellular mass remains in the ovary after the egg is released. It is called the **corpus luteum**. There is a fatty substance stored in the corpus luteum that contains the hormones **estrogen** and **progesterone**. These hormones are released into the bloodstream and stimulate the growth of the inner uterine wall (**endometrium**) for possible implantation of an embryo.

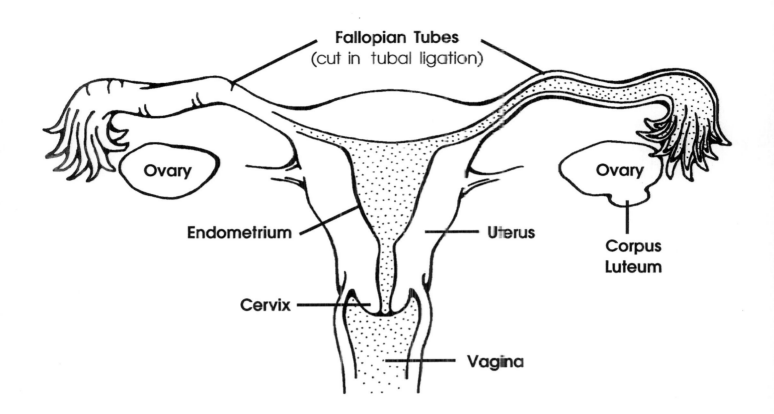

The monthly menstrual cycle is fairly complex but can be summarized as:

Day 1   Menstrual flow begins from previous month's cycle. One egg begins to mature in the ovary for release in the next cycle.

Day 14–15   Ovulation. (Ovary releases an egg.) The corpus luteum produces hormones that stimulate the endometrium of the uterus to grow.

Day 28   If the egg isn't fertilized, then the uterus lining is shed with the menstrual flow.

? QUESTION

1. Health officials warn: "The eggs of your daughters can be damaged by certain chemicals you are exposed to during pregnancy." How can the eggs of your daughter be damaged when she won't start menstruating until 12 or 13 years after her birth?

2. Which endocrine gland produces the hormone that stimulates an egg to mature?

3. Where does the corpus luteum form, and what hormones does it produce?

4. What day of the menstrual cycle does ovulation usually happen?

5. The menstrual flow is a shedding of the _____ lining.

| GO GET |

1. A compound microscope.

2. Three microscope slides: Immature Egg
                                  Mature Egg
                                  Corpus Luteum

| NOW |

1. Work with two other lab groups. Put a different slide on each microscope, and use low magnification to get an overview of the sectioned ovaries.

2. The immature eggs are bigger than most other cells in the ovary. A maturing egg cell is larger and has a ring of cells surrounding it. The spherical mass of cells is called the follicle. The corpus luteum is even larger than the follicle, and is a solid mass of cells without an egg inside. (The egg has been released.)

3. Look back and forth among the three slides until you can see the difference in the structures.

4. Draw each structural stage in enough detail so that you can find it again (perhaps on a test) with the help of your picture.

Immature Egg   Mature Egg   Corpus Luteum

5. *Return the slides.*

SPERM

The reproductive system of the human male embryo is noticeably differentiated by the eighth week. However, the male fetus does not begin meiotic division of sex cells as does the female fetus. The female has immature eggs that are stopped at Prophase I of meiosis. The male doesn't begin sperm production until puberty.

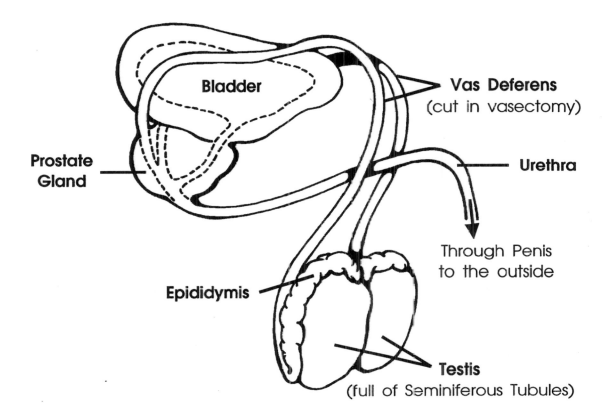

It takes about three weeks for sperm to be produced. If a male is exposed to damaging chemicals, only the current "batch" of sperm is affected. After a month, sperm production could be back to normal. Of course, chromosome damage does sometimes happen in the male gametes as in the female. But, consider how little time sperm chromosomes are exposed to potential damage by environmental factors compared to egg chromosomes.

Sperm are produced (200–500 million daily) along the inside of an extensive system of tubes (**seminiferous tubules**) in the testes. They are stored in enlarged tubes called the *epididymis*.

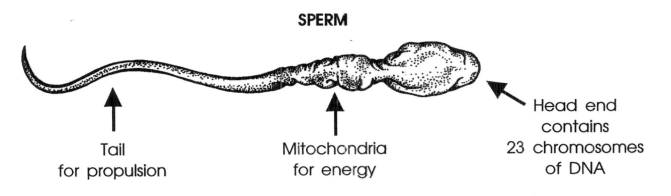

313

1. Explain how sperm are less vulnerable to chromosome damage than eggs.

2. Where specifically are sperm produced?

3. Where are they stored?

GO GET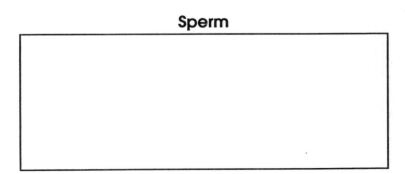

Two microscope slides: Testis
                          Human Sperm

NOW

1. Examine the cross-section of a testis, and find the seminferous tubules filled with various stages of sperm production.

2. Draw a picture of the testis. Label the seminiferous tubules and sperm.

3. Examine the slide of human sperm. Identify the head, mitochondrial region, and tail. Make a sketch and label it.

4. *Return the slides.*

Testis

Sperm

FERTILIZATION

The human egg is released from the ovary with a layer of follicle cells stuck to its outside. Sperm cannot fertilize the egg until this layer of cells has been removed. Each sperm cell produces a small amount of enzyme that dissolves the jelly-like glue holding the small cells to the egg. Much of this enzyme is necessary to separate the follicle layer from the outside of the egg.

Eventually, enough of the protective layer surrounding the egg is dissolved, and one of the sperm penetrates the egg. Once that happens, the egg membrane reacts by forming a new layer of jelly around the egg, thereby preventing any more sperm from entering it. If more than one sperm fertilizes an egg, the resulting embryo dies.

Fertilization usually occurs as the egg moves through the fallopian tube towards the uterus. The embryo continues to develop for about one week before it implants in the wall of the uterus. After implantation, a hormone from the growing embryo signals the corpus luteum (in the ovary) to produce large amounts of estrogen, which then stimulates the mother's reproductive system to prepare for pregnancy. Estrogen also prevents new eggs from developing during pregnancy. (This is why estrogen can be used for birth control.)

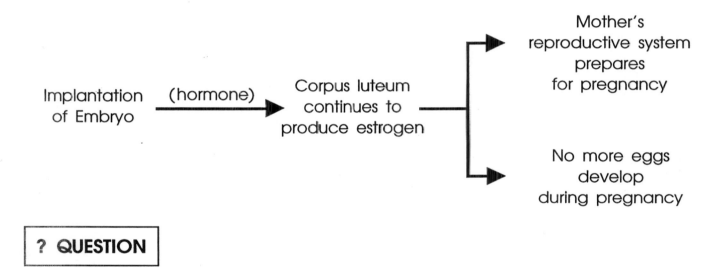

? QUESTION

1. It is necessary for a male to have a high sperm count to be fertile. If only one sperm is necessary to fertilize the egg, then what function is served by the other sperm?

2. What event signals the mother's reproductive system to prepare for pregnancy?

3. What prevents more than one sperm from fertilizing the egg?

A depression slide and coverslip.

NOW

If your instructor has been able to get mature sea urchins, then this is the time to watch fertilization actually happen.

1. Your instructor will inject the urchins with a salt solution that stimulates the release of their gametes. The urchins are then placed upside-down in a beaker of sea water. The gametes will flow out of the animal.

2. Sperm looks milky. The eggs are granular and usually have a slight pink or yellow color.

3. Put a drop of the eggs into a depression slide (no coverslip). Carefully place the slide on your microscope and examine the unfertilized eggs using the 10x objective lens.

4. The sperm must be diluted with sea water because too many sperm causes abnormal fertilization. If the sperm collection beaker looks slightly milky, then dilute the sperm further. Add a drop of the sperm to your egg slide. Cover the slide with a coverslip, and immediately observe the events under the microscope. A *fertilization membrane* usually forms within 2 minutes.

5. When you've seen the membrane form, put this slide aside, and recheck it every 30 minutes. Don't leave the slide on the microscope with the light on because fertilized eggs will soon overheat. If everything goes well, the first *cleavage* (division) should happen in about an hour. Watch for it!

6. Draw pictures of the following:

Unfertilized Egg	**Fertilization Membrane**	**Cleavage**

ACTIVITY #3

"METHODS OF FEEDING THE EMBRYO"

QUICK-DEVELOPMENT SELF-FEEDING

Invertebrates usually give no special care to their embryos. The only source of food during the first stages of development is provided by the cytoplasm in their eggs. After fertilization, there is rapid cell division and development to allow the embryo to feed on its own. As one example, the sea urchin embryo develops to a self-feeding stage in 24 hours or less. There are many small pieces of organic matter and micro-organisms in water for young embryos to eat. (You will see sea urchin embryos during Activity #4.)

YOLK

Nutrition in the form of *yolk* is another way of giving food to the embryo. Fish are examples of organisms that have small amounts of yolk in the egg. Birds and reptiles have much more yolk in their eggs. Larger amounts of yolk provide longer possible times for development. This is especially necessary for the more advanced land vertebrates.

**Sea Urchin
& Human
Egg Cell**
(very little yolk)

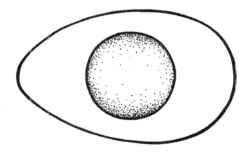

O

**Frog
Egg Cell**
(a little more yolk)

**Reptile & Bird
Egg Cell**
(very large yolk)

PLACENTA

Mammals provide nutrition to their embryos through a specialized integrated feeding structure called the **placenta**. It is formed partly by the embryo and partly by the mother. Nutrients, wastes, and blood gases are exchanged between the embryo's blood system and the mother's blood system. The placenta begins as a few specialized cells on the outside of the implanting embryo. These special cells digest their way into the wall of the uterus. Soon after implantation, finger-like projections begin the formation of a more efficient structure. By the second month, the placenta is well developed, and continues to grow with the fetus.

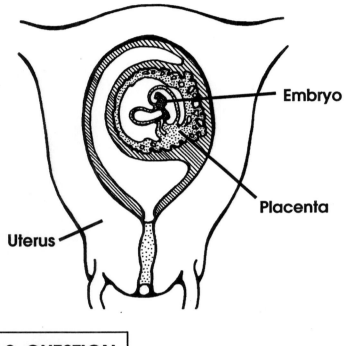

The human embryo, from the zygote to blastocyst stage, must survive on the cytoplasm of the original egg cell. The egg divides but cells don't grow in size. In about one week the developing human embryo begins to implant into the uterine wall for the next stage of feeding (the placenta).

? QUESTION

1. List three methods for feeding early embryos by different organisms.

   a.

   b.

   c.

2. Why do complex animals require more specialized embryonic feeding mechanisms?

3. Why do most land animals require a more developed embryonic feeding mechanism than aquatic animals?

ACTIVITY #4

"STAGES OF DEVELOPMENT"

During this Activity, you will examine 10 stages of vertebrate development using various organisms from sea urchin to pig embryos to illustrate each stage. We will skip many of the details, and focus only on a few structures as examples of the events.

CLEAVAGE, MORULA, AND BLASTOCYST

The first divisions in development are called **cleavage** because the zygote divides (or cleaves) into two cells, then four, then eight, etc. Eventually, a small solid ball of cells forms called the **morula**. Cells of the morula continue to divide, producing a hollow ball of cells called the **blastocyst**. The human blastocyst develops in about one week after fertilization. You can get an idea of what these early stages look like by observing early embryos of sea urchin and frog.

Two microscope slides: Early Sea Urchin Development
Frog Blastula (stage 8)

NOW

1. Examine the early developmental stages of the sea urchin.

2. Compare the sizes of the earliest embryos undergoing cleavage. Are the stages from zygote to morula about the same size? _____ Are these early cells growing or just dividing? _____

3. Draw pictures of these stages:

Zygote	Early Cleavage	Morula

Frog Blastocyst

1. Examine the early frog embryo.

2. The frog blastocyst (also called *blastula*) is a partial hollow ball of cells. You should be able to see several different cell types. These cells are changing into new kinds of cells for the next stage of development.

3. Draw a picture of the frog blastocyst, and show the different cell types.

4. *Return the slides.*

EARLY CHICK EMBRYOS

GO GET

Two microscope slides: 21-Hour Chick Embryo
                          28-Hour Chick Embryo

NOW

1. Work with another lab group. Put the two slides on different microscopes so that you can look back and forth between them.

2. The 21-hour embryo shows early development of the nervous system. The long **neural fold** becomes the spinal cord in a later stage. Perhaps you can see that the head end is slightly larger.

3. The 28-hour embryo has a more developed neural fold, and you can see that the brain end has enlarged. Also, there are several paired blocks of tissue along the **neural tube**. These blocks are called **somites**. The somites develop into the internal organs, skeleton, and muscles of the adult body.

4. Draw pictures of the two embryonic stages, and label the neural fold, neural tube, head end, and somites.

| 21-Hour Chick Embryo | 28-Hour Chick Embryo |

5. *Return the slides.*

LATER CHICK EMBRYOS

Two microscope slides: 56-Hour Chick Embryo
72-Hour Chick Embryo

NOW

These chick embryo stages are comparable to three and four-week-old human embryos.

1. Put the two slides on different microscopes for comparison. These are side views of the embryos, which are shaped like a question mark (either forward or reversed depending on how the slide was made).

2. You should be able to see changes in the brain development between the two embryonic stages. The brain consists of several lobes. In the 72-hour stage there is a lobe in front of the lobe with the eye in it. That front lobe is the **cerebrum**.

3. The somites are easy to see. How many are there? _____

4. There is a structure bulging from the middle of the embryo, and positioned below the brain. This is the developing heart. In the older embryo, the bottom chamber of the heart is larger. That larger chamber is the *ventricle*.

5. Draw a picture of the 72-hour embryo showing brain and spinal cord, eye, heart, and somites. Label your drawing.

6. *Return the slides.*

72-Hour Chick Embryo

GO GET

The slide of the 10-mm pig embryo.

NOW

The 10-mm pig embryo is comparable to a human embryo stage at six to eight weeks.

1. You should recognize a well developed nervous system, including backbone, large brain, but smaller eyes than in the chick embryo.

2. Can you find the *limb buds*? In the human embryo these buds will develop into arms and legs.

3. Just below the heart is a large dark structure, the *liver*.

4. Draw a picture of the pig embryo showing brain, eye, vertebral column, heart, liver, and limb buds. Label your drawing.

5. *Return the slides.*

10-mm Pig Embryo

ACTIVITY #5

"DISRUPTION OF NORMAL DEVELOPMENT"

Considering all the events that must go exactly right during embryonic development, it is easy to see that events can go wrong. The general principle for understanding developmental disorders is:

The earlier the disruption, the more serious are the consequences.

Some researchers estimate that one-third of conceptions do not make it to birth. Women are usually unaware of the earliest embryonic failures, which are also the most common. However, there are several disruptions of later development that are very serious and reveal themselves at birth (4–5% of births), or as late-term miscarriages.

GENE MUTATIONS

Gene mutations can destroy the normal production of essential enzymes controlling basic metabolic processes in the fetus.

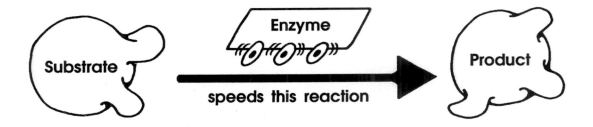

Most states require the immediate testing of all babies for PKU (phenylketonuria), which is a deficiency of an enzyme responsible for the metabolism of the amino acid *phenylalanine*. This disorder is similar to other gene mutations—no product is formed, and there is a buildup of the substrate. High concentrations of these substrates are usually toxic or disruptive, and often the product is essential for other processes. Tay Sachs, galactosemia, and tyrosinosis are other metabolic disorders resulting from gene mutations.

CHROMOSOMAL ABERRATIONS

When mistakes occur during meiosis of sperm or egg production, abnormal numbers of chromosomes in the embryo are possible. These events are almost always lethal. However, some chromosomal aberrations (X, XXY, and three #21 chromosomes) occur in about 1% of births. These babies have serious health problems. The *trisomy* of #21 chromosomes produces Down's syndrome, the most common form of mental retardation. The other chromosomal aberrations also result in similar problems.

ECTOPIC PREGNANCY

Occasionally, the embryo implants in the wall of a fallopian tube, or on the outside of the uterus or intestine. There are two serious consequences of this abnormal event. First, the child will have to be delivered by Cesarean section. The second serious problem is that the placenta can attach to the wall of the wrong organ. That organ may be damaged or destroyed by placental development. This condition can be so dangerous to the mother's life, that termination of pregnancy may be her only chance of survival.

POSSIBLE SITES OF IMPLANTATION

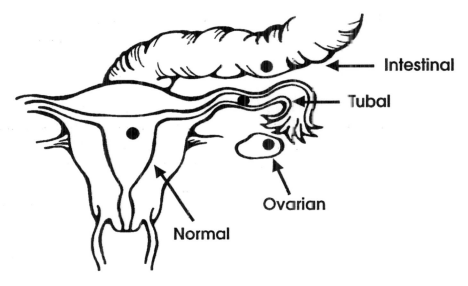

Intestinal

Tubal

Ovarian

Normal

INCOMPLETE DEVELOPMENT OF ORGAN

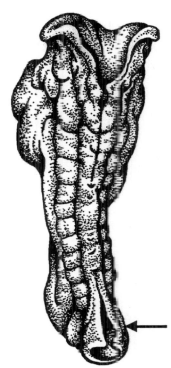

Sometimes an organ does not develop completely during early development. The cause can be a foreign chemical like thalidomide, or a disease like measles. Common examples are spina bifida (neural tube doesn't close), or holes between the heart chambers.

The general principle of developmental disorders applies especially in these cases. The earlier the problem happens, the more serious are the consequences.

Spina bifida occurs if this part of the neural tube does not close during early development.

325

BAD HABITS DURING PREGNANCY

We have very little control over the previously described disruptions of normal development. In fact, many early miscarriages are the result of the mother's hormonal system "warming up" for future reproductive success.

However, there are disruptions in development that we *can* control—bad habits during pregnancy. Science makes a certain direct statement about bad habits during pregnancy:

"Drink alcohol, smoke cigarettes or marijuana, use drugs, or have poor nutrition, and your child will be below normal in mental abilities, and may be physically deformed."

FETAL ALCOHOL/DRUG SYNDROME

This fetal damage is predictable and results from the mother's drinking or drug habit. Fetal Alcohol/ Drug Syndrome includes characteristic facial deformations, a degree of mental retardation, and defects in internal organs (especially the heart and liver).

CIGARETTE SMOKING

The majority of babies born by mothers who smoke, have a smaller placenta, lower birth weight, and lower overall health. The effects on the fetus are not usually as extreme as with alcohol or drug use by the mother, but the birth results are 100% predictable.

POOR NUTRITION

The usual problem with poor nutrition in this country is not a lack of Calories, but an unbalanced diet (not enough protein, vitamins, and minerals). The effects of poor nutrition are lower birth weight, lower overall health, and below-average mental ability.

If the nutritional problems during pregnancy are severe, as seen in overpopulated parts of the world, all of the detrimental effects during embryonic and fetal development are greatly magnified.

1. What is the general principle for understanding developmental disorders?

2. Perhaps as many as _____ of all embryos don't survive to birth.

3. What is the percent of births with developmental defects?

4. What is the most serious potential risk in an ectopic pregnancy?

5. List the two immediate biochemical effects of most gene mutations.

   a.

   b.

6. What is the usual result of chromosomal aberrations?

7. What are the common effects on embryonic development caused by alcohol, drugs, cigarettes, and poor nutrition?

Summary Questions

1. What is a circadian rhythm?

2. What sets a "biological clock?"

3. What term is used when there is no obvious circadian rhythm or when the separate rhythms do not rise and fall together?

4. What is the most probable cause for question #3 to occur in college students?

5. List some possible health consequences of abnormal circadian rhythm.

6. List some possible occupational consequences of abnormal circadian rhythm.

7. If disrupted circadian rhythm continues over a long period of time, or as a lifestyle pattern, what could be the serious consequence?

CIRCADIAN RHYTHMS

INTRODUCTION

Life is exposed to major cycles of environmental change. Obvious examples are the day-night cycle, the annual cycle of seasons, and lunar and tidal cycles. Organisms must adjust their activities to these rhythmic changes in nature. Plant and animal physiologists have verified that planetary cycles influence the biochemistry of organisms. Do humans have internal cycles?

During this week's lab you will investigate one category of cyclic change in your body—the *circadian rhythm*. These are daily rhythmic changes in behavior or physiological processes (circadian = about a day).

You will monitor changes in your heart rate, body temperature, and reaction time during three days. In addition, you will collect data about nose breathing cycles, and keep a sleep diary for one week. To get the best understanding of your own personal physiology, be accurate and do all of the activities as directed. This will be a busy week of data collection.

ACTIVITIES

ACTIVITY#1

"FILM—*WHAT TIME IS YOUR BODY?*"

You will watch a short film that summarizes some of what is known about circadian rhythms. Read the following questions now, and answer them as topics are presented in the film.

? QUESTION

1. Define *circadian rhythm.*

2. Define "free-running" rhythm.

3. Define *desynchronization* of rhythms.

4. What "sets" a biological clock?

5. Is the timing of a circadian rhythm firmly established at birth, or can it be changed?

6. List the evidence supporting the idea that circadian rhythms occur in most life-forms.

7. List the evidence supporting the idea that circadian rhythms are related to health.

ACTIVITY #2

"INVESTIGATING YOUR CIRCADIAN RHYTHMS"

Careful observations during this week can reveal interesting and useful insights about your physiology. Good luck, and have fun (even at 3 AM).

HEART RATE, BODY TEMPERATURE, AND REACTION TIME

Pick three consecutive days during the week to monitor your heart rate, body temperature, and reaction time. Measure these physiological parameters six times each day as close as possible to the indicated hours. Record the data in the Circadian Rhythm Chart on the next page.

Heart Rate Tips: Before counting heart beats, sit quietly and relax for 2–3 minutes. Count your heart beats (wrist pulse) for 30 seconds, then convert that number to the rate per minute.

Body Temperature Tips: Use an oral thermometer to measure body temperature. Do not drink anything hot or cold 15 minutes beforehand.

Reaction Time Tips: An easy-to-perform test of your reaction time is finger counting. Determine the amount of time (in seconds) it takes to do 10 finger-counts on one hand. Do a single finger-count in the following manner:

> Touch your thumb to your first finger,
> then second finger, third, fourth; fourth
> finger again, third, second, and finally
> first finger again.
>
> That's one finger-count.
> Now, do ten counts.

Warm up your reflex system by practicing finger counting several times before you actually do each test during the day.

Use a watch with a second hand, do 10 finger-counts as fast as you can, and record the elapsed time (in seconds).

CIRCADIAN RHYTHM CHART

		3 AM	7 AM	11 AM	3 PM	7 PM	11 PM
Heart Rate (beats per minute)	Day 1						
	Day 2						
	Day 3						
Body Temperature (°F)	Day 1						
	Day 2						
	Day 3						
Reaction Time (in seconds)	Day 1						
	Day 2						
	Day 3						

OPEN NOSTRIL TEST

Do these tests during 6 consecutive hours of any one day during the week.

At the beginning of each hour, breathe in while first closing one nostril and then the other. Determine which nostril is the primary one you are breathing through. Give each nostril a rating: "open," or "closed," or "both" (if both nostrils are mostly open at the same time).

OPEN NOSTRIL TEST		
Time	Right Nostril	Left Nostril
Hour 1		
Hour 2		
Hour 3		
Hour 4		
Hour 5		
Hour 6		

SLEEPING PATTERNS

Record your sleeping patterns during the week. Complete the sleeping pattern chart in Activity #3. Present the data in the following manner:

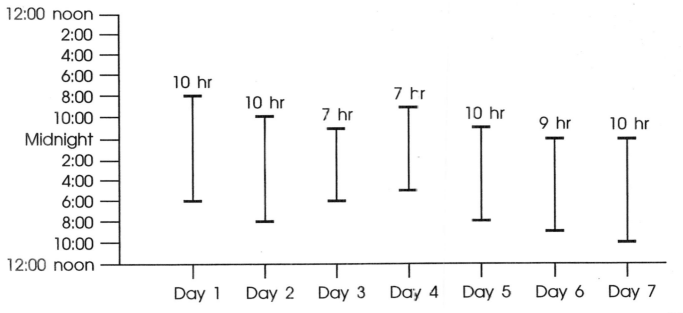

MORNING PERSON OR EVENING PERSON

Answer the questionnaire, and determine your morning-person/evening-person score. (Add the number values of the answers that you give to the ten questions.)

QUESTIONNAIRE

Directions: Answer these questions in order. Circle your choice. Do not go back and change your answers. Think for awhile about each question before answering it so that you make your most accurate first choice.

1. At what time in the evening do you notice that you are getting sleepy?
   1) 8 PM
   2) 9 PM
   3) 10 PM
   4) 11 PM

2. How alert do you feel during the first hour after waking up in the morning?
   1) very alert
   2) fairly alert
   3) slightly alert
   4) not at all alert

3. If there were no other factors to consider, when would you like to eat your biggest meal?
   1) breakfast
   2) lunch
   3) mid-afternoon dinner
   4) evening dinner

4. When would you prefer to get up in the morning?
   1) 6 AM
   2) 7 AM
   3) 8 AM
   4) 9 AM

5. When would you prefer to go to bed in the evening?
   1) 9 PM
   2) 10 PM
   3) 11 PM
   4) 12 PM

6. To what degree are you dependent on an alarm clock to get up in the morning?
   1) not at all dependent
   2) slightly dependent
   3) fairly dependent
   4) very dependent

7. If you had no commitments tomorrow, what time would you go to bed tonight (compared to normal)?

   1) same time
   2) within 1 hour later
   3) 1–2 hours later
   4) more than 2 hours later

8. Assume that you have to do four hours of hard physical work. When would you choose to do this work if you had a choice?

   1) 6 AM to 10 AM
   2) 8 AM to noon
   3) 10 AM to 2 PM
   4) noon to 4 PM

9. Assume that you have a very important two-hour-long test to take. When would you choose to take this test if you had a choice?

   1) 8 AM
   2) 11 AM
   3) 2 PM
   4) 5 PM

10. In terms of being a morning or evening person, how do you generally describe yourself to other people?

    1) definitely morning
    2) somewhat morning
    3) somewhat evening
    4) definitely evening

Add the number values of all your ten answers, and record the total here:

Total = _____

A score of 10–20 = morning person

A score of 30–40 = evening person

ACTIVITY #3

"ANALYZING THE DATA"

Complete the following graphs using your personal data.

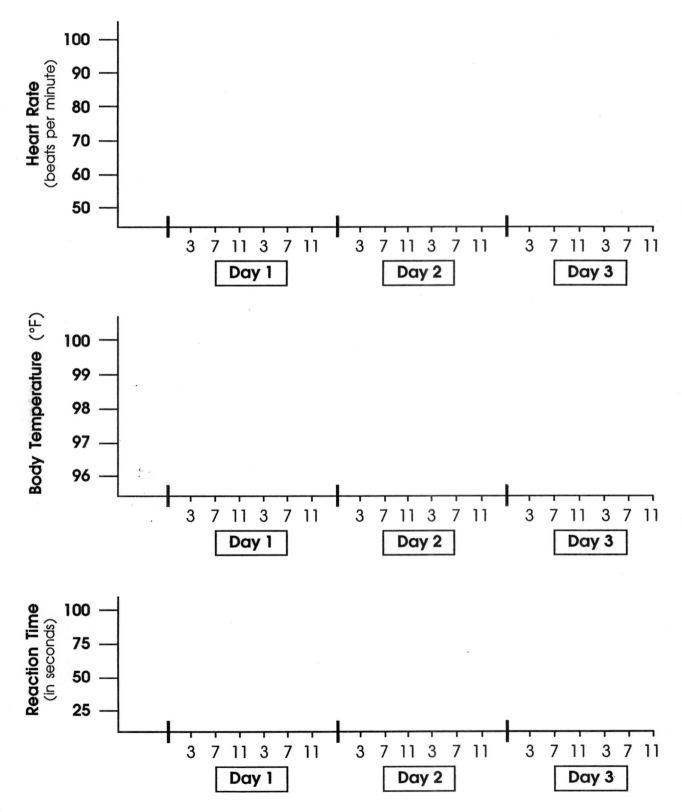

Calculate the average time of day that your heart rate, body temperature, and reactivity were at maximum and minimum in their cycles. Record your findings.

CYCLE		AVERAGE TIME OF DAY
Heart Rate	Highest	
	Lowest	
Body Temperature	Highest	
	Lowest	
Reaction Time	Fastest	
	Slowest	

Use the class data to make bar graphs showing the number of students who were "peaking" or "ebbing" at each of the times of the day.

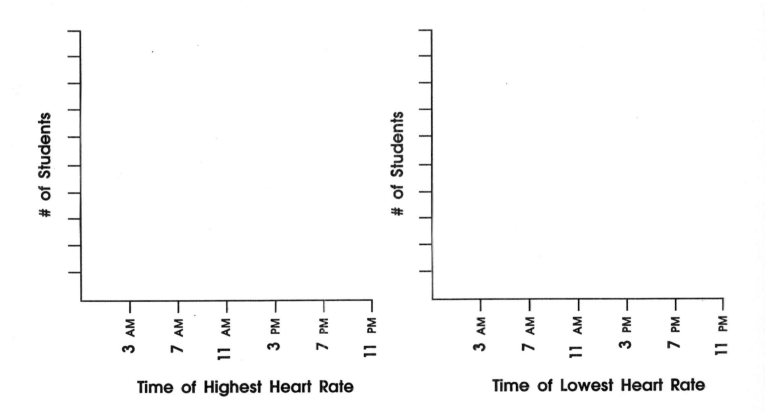

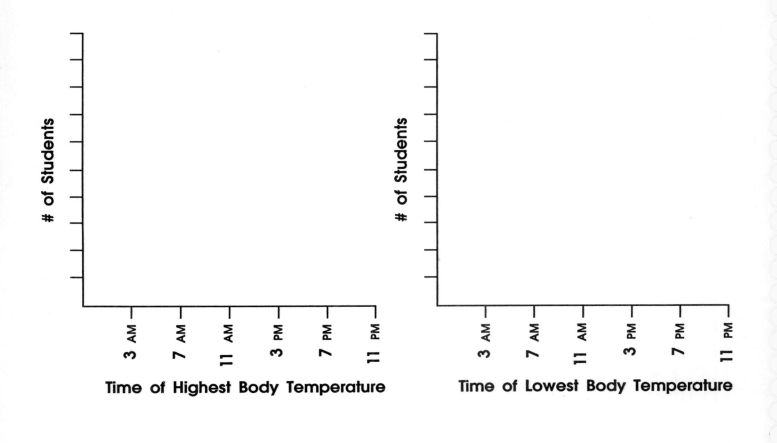

# of Students

Time of Highest Body Temperature

3 AM 7 AM 11 AM 3 PM 7 PM 11 PM

# of Students

Time of Lowest Body Temperature

3 AM 7 AM 11 AM 3 PM 7 PM 11 PM

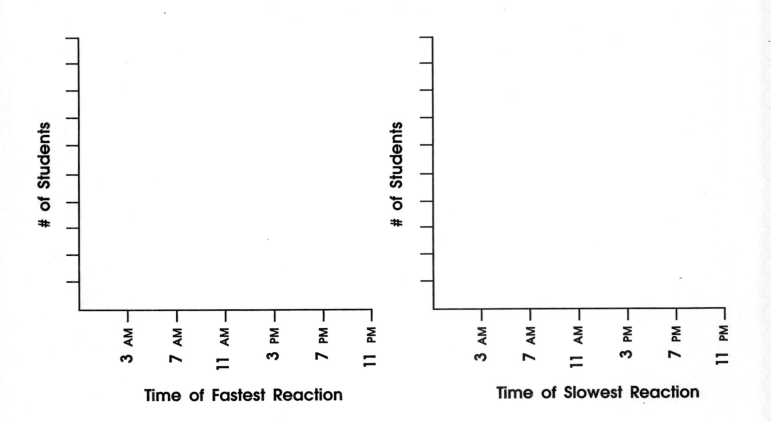

# of Students

Time of Fastest Reaction

3 AM 7 AM 11 AM 3 PM 7 PM 11 PM

# of Students

Time of Slowest Reaction

3 AM 7 AM 11 AM 3 PM 7 PM 11 PM

YOUR SLEEPING PATTERNS

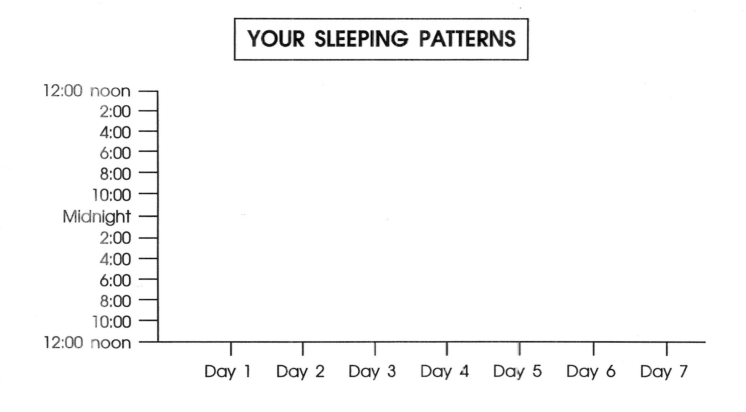

? QUESTION

1. Researchers have concluded that well-defined circadian rhythms in heart rate and body temperature are indicators that a person has an efficiently timed physiology. It is common for college students to show disrupted rhythms.

   Do you have well-defined rhythms in heart rate and body temperature? _____

   What percent of your classmates have well-defined rhythms? _____

2. Look for examples of both well-defined and disrupted rhythms among your classmates. Determine if the people with well-defined patterns are less sleepy during the day than other students.

   Were they less sleepy? _____

3. Some investigators think that you should study when your reaction time is in its fastest phase of the day. What time of day was your reaction time the fastest? _____

Do you usually study during that time? _____

4. Medical research suggests that desynchronized rhythms (body temperature peaks at different time than heart rate) occur as a result of illness. Find someone in the class who has been ill recently, and determine if they have desynchronized rhythms.

Did they? _____ How long ago were they ill? _____

5. What is your morning/evening score? _____

Does your morning/evening score correlate with the time that most of your physiological rhythms are at their peak? _____

What generalization is apparent among the class data?

6. Did you notice a definite pattern of nostril opening and closing? _____
Describe your pattern.

7. Each nostril is on its own cycle. A nostril is open for awhile, and air flows easily through it. However, the air begins to irritate the nostril. It swells and begins to close. This causes the air flow to decrease, irritation is reduced, and the nostril recovers.

Knowing this, should you spray nasal decongestant into both nostrils when you have a stuffy nose? _____ Explain.

8. Sleep research indicates that there is a range in the number of sleep hours necessary for good health. For some, six hours of sleep are necessary, and for others nine hours are necessary. What was the range of average sleeping hours in your class?

Most = _____ Least = _____ Yours = _____

9. Our brains don't consciously keep track of sleeping patterns (sort of like eating). Survey the students with irregular sleeping hours and find out if they were surprised to discover their actual sleep habits.

   What percent of these students were surprised? _____

10. Irregular sleeping causes increased complaints of sluggishness during the day and more frequent illness. A change of only two hours in the normal sleeping schedule on more than two days of the week correlates with health complaints.

    Using this standard, do you have a disrupted sleep schedule? _____

    On what days? _____

    What percent of the class has disrupted sleep schedules? _____

SUMMARY

Researchers were very much interested in human circadian rhythms during the 1970s. Studies indicated that recovery from surgery, response to medications, accidents, and best performance followed circadian rhythms. Other research revealed dramatic effects because of irregular work and sleep schedules. There could be as much as a 25% decrease in lifespan for those people who regularly rotate shifts during their work careers.

The interest in circadian rhythms has declined as researchers discovered that both individuals and society were reluctant to change life-style patterns. However, opportunities for a longer and healthier life still exist for anyone interested in the subject. All you have to do is change a few habits.

How about it?

344

Summary Questions

1. Do identical twins have idential fingerprints?

   How about their DNA fingerprints?

2. What is PCR?

3. Restriction enzymes cut what molecule?

4. How do the recognition sites (for restriction enzymes) differ from person to person?

5. Define gel electrophoresis.

6. What is the purpose of a genetic probe (either radioactive or fluorescent)?

7. Define recombinant DNA.

BIOTECHNOLOGY: DNA

INTRODUCTION

Biotechnology uses organisms and their biological processes to produce new solutions for medical, agricultural, and commercial problems, as well as many other social and environmental applications.

This research area is advancing rapidly and its applications are expanding comtemporary imagination. Its advances can be compared to the invention of the computer with the blossoming information age. Biotechnology is fueled by discoveries of the biochemical events both controlling and controlled by biomolecules, including the structure and function of proteins (enzymes), RNA, and DNA. It is as if we have discovered a basic secret of life—all that it creates, and that it could create.

Has biotechnology uncorked the bottle and let out the Genie?
For what do we wish?
Which wishes will improve the world or human condition?
How will we avoid making a foolhardy wish?
This lab introduces a small part of the biotechnology world, focusing on DNA technologies and how they are valuable to us.

ACTIVITIES

ACTIVITY #1

"DNA FINGERPRINTING TOOLS"

DNA fingerprinting (more accurately called DNA profiling) is a procedure used to identify specific characteristics in DNA molecules. This procedure allows us to distinguish the DNA of different people or of any other organism. DNA fingerprinting has been applied to criminal investigations, paternity cases, genetic-relationship questions, identification of inherited disorders, and the personal identification of individual humans.

The classic use of fingerprints taken from the hand is for identification purposes only, and it lacks the ability to show genetic relationship between people. Identical twins have different hand fingerprints because those prints are partly determined by non-genetic conditions during embryonic development. Whereas classic fingerprints focus on whorls and intersections of ridges on the fingers that are not inherited, DNA fingerprinting identifies actual sequences of nucleotides in the DNA. DNA fingerprinting uses many biotechnology tools, and a discussion of several of those tools will give you a basic understanding of the process.

Classic Fingerprint "markers"   **vs**   **DNA Fingerprint "markers"**

TOOL 1: PCR—HOW TO CLONE DNA

PCR (polymerase chain reaction) is a method used to make millions of copies of DNA from a small sample. It is sometimes called "molecular photocopying." The DNA is "unzipped" by heating the sample in a mixture of lab-made DNA nucleotides and a special enzyme called *Taq DNA polymerase*. A small amount of "primer" (short piece of RNA that starts the chain reaction) is also added to the mixture. The DNA molecule copies itself (A pairs with T; G pairs with C) under these conditions.

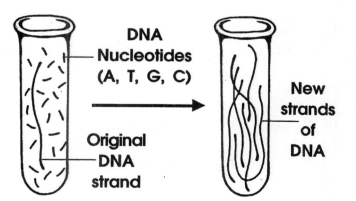

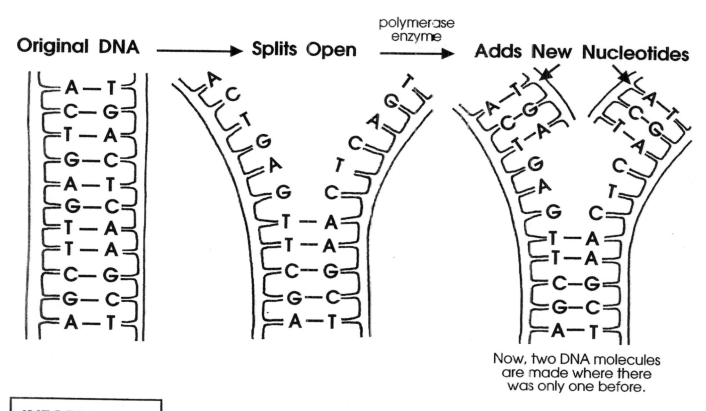

Original DNA → **Splits Open** —polymerase enzyme→ **Adds New Nucleotides**

Now, two DNA molecules are made where there was only one before.

INFORMATION

The special Taq polymerase was discovered and extracted from thermal-pool bacteria like those in Yellowstone National Park. The bacteria was named *Thermus aquaticus*, hence the name Taq polymerase. This enzyme is capable of functioning at the high temperatures of the laboratory PCR process. Most enzymes are destroyed by high heat. Use of this enzyme is another example of a biotechnology advancement.

There are a couple of tricks to this procedure. The starting sample is heated to about 95 °C. This causes the DNA molecule to unzip into two complementary halves. When the DNA is cooled to about 50 °C, each half makes a new *complementary* copy of itself from the mixture of nucleotides. Heating again separates the two new DNA molecules and another replication occurs when the sample is cooled for a second time. An automated machine can repeat these temperature change cycles (30–40 times) to produce millions of exact copies of the original DNA. The PCR technique creates plenty of sample for the DNA technician to do specific DNA tests.

? QUESTION

1. What is a simple definition of biotechnology?

2. In which example is the "fingerprint" identical for identical twins? (circle your choice)

        Classic hand fingerprints   or   DNA fingerprints

3. What does PCR produce?

4. What are the ingredients for the PCR process?

5. Where did biotechnology find the special DNA polymerase enzyme used in the PCR process?

TOOL 2: RESTRICTION ENZYMES

When long molecules of DNA are mixed with **restriction enzymes**, the DNA is cut into shorter threads. Restriction enzymes act like "scissors" cutting the DNA at specific nucleotide sequences called **recognition sites**. Let's see if you get the idea of how this biotechnology tool is used.

Start with three examples of restriction enzymes (BamHI, EcoRI, and HindIII). Each enzyme cuts at its own recognition site. A recognition site could be as simple as CCCGGG where the restriction enzyme cuts the DNA between the G and C. There are hundreds of restriction enzymes developed by biotechnology and each has its own technical name. We have used a simple shape to represent the recognition site for each of the three cutting enzymes above. The particular restriction enzyme cuts only where that shape occurs in the long thread of DNA.

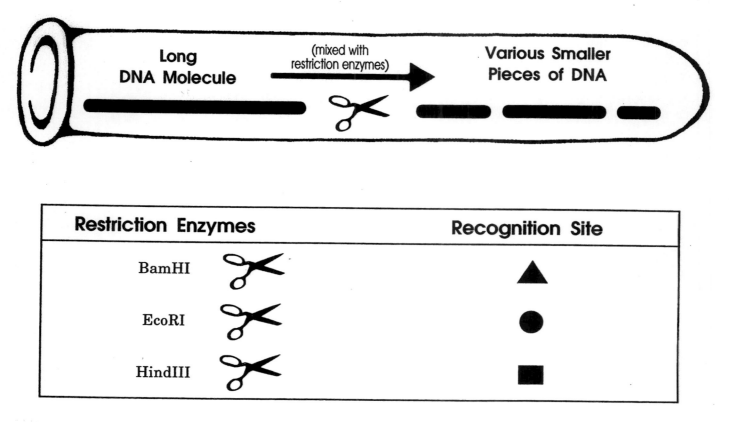

1. Your job is to analyze the DNA of three subjects. Their DNA was mixed with the three restriction enzymes ("scissors").

2. The long threads of DNA from each subject are cut at the recognition sites. Determine the length of each resulting piece of DNA from the subjects after cutting has occurred.

3. Measure the length of each piece with a cm ruler and make a thick line mark on the graph to indicate each of the pieces.

4. We have done this for the first piece in Subject A. Finish measuring and making marks for all the DNA pieces of Subjects A, B, and C.

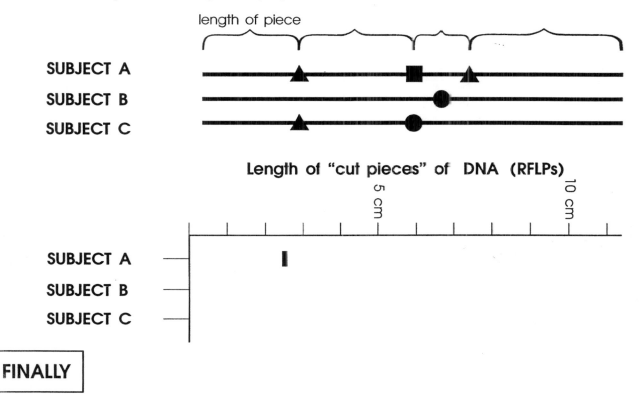

FINALLY

You now have a visual picture of a DNA comparison of Subjects A, B, and C.

INFORMATION

1. The scientific name for these individual pieces of DNA cut up by restriction enzymes is **RFLPs** (Restriction Fragment Length Polymorphism).

2. RFLPs are separated from each other during actual DNA fingerprinting in the process called *gel electrophoresis*, which is the biotechnology tool described next.

3. The length and number of RFLPs is dependent on the restriction enzymes used and the unique DNA code in each person.

4. The forensic use of DNA fingerprinting has particular criteria for selecting restriction enzymes to analyze DNA crime scene samples.

TOOL 3: GEL ELECTROPHORESIS

Gel electrophoresis is a precise method for separating pieces of DNA (called RFLPs) that have been cut by restriction enzymes. It is the most familiar part of the DNA fingerprinting process because it produces a visual gel record similar to the graph in the previous discussion.

INFORMATION

1. PCR amplifies the original DNA sample.

2. Restriction enzymes cut the DNA into smaller pieces.

3. DNA pieces are separated from each other by gel electrophoresis.

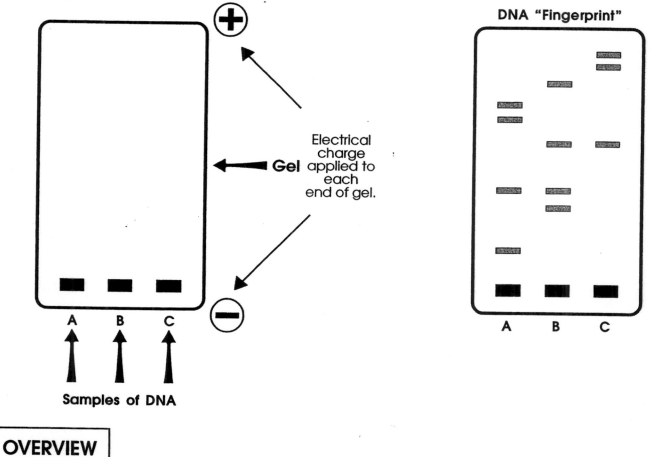

Electrical charge applied to each end of gel.

Gel

DNA "Fingerprint"

Samples of DNA

OVERVIEW

1. ***Prepare an agarose gel sheet in a casting tray.*** Agarose is made from seaweed. This gel-like sheet provides the medium through which RFLPs move. The structure of this gel allows different size pieces of DNA to separate from each other. Think of the gel as a matrix of many small particles suspended in it. The particles form an "obstacle-course" with uniform size openings between the particles. Bigger pieces of DNA move slower through the gel than smaller pieces.

2. ***Cover the gel sheet with a buffer solution.*** This stabilizes the pH and serves as a conductor of electricity between the gel and the electrodes during the electrophoresis.

3. *Place the samples to be compared into small "wells" at one end of the gel sheet.*

4. *Apply an electrical charge to each end of the gel.* DNA pieces (which have a negative charge) begin to move away from the negative electrode and towards the positive electrode.

5. *Wait for the "DNA fingerprint" to develop.* This might take 30 minutes or so.

This basic technique can be modified to improve the separation of DNA fragments. RFLPs of identical size but different genetic coding (for example, different GC content) will separate into multiple bands in the gel.

? QUESTION

1. What are restriction enzymes?

2. What are recognition sites?

3. What are RFLPs?

4. What is the purpose of the gel?

5. What is the purpose of the buffer?

6. Where do you put the DNA samples?

7. When the electrical charge is applied to the gel, what happens?

8. Why are there different "bands" when the process is finished?

TOOL 4: RADIOACTIVE AND FLUORESCENT PROBES

Radioactive or *fluorescent probes* are special molecules that are designed to combine only with a particular sequence of nucleotides in the DNA sample. The "targeted" sequences are called *genetic markers*. When the marker has combined with a specific probe, that DNA is either radioactive or fluoresces when a special UV light is shined on it. Researchers produce targeting probes to combine with a particular gene, or with a specific gene variation responsible for disease. There are many useful applications of this biotechnology tool.

OVERVIEW

1. Design an appropriate probe to target the genetic marker.

2. Add the probe to the DNA sample you want to test, and incubate at the correct temperature for the probe.

3. Wash the sample. This removes the probe if it has not combined with a targeted gene in the DNA.

4. Test the DNA sample to see if it is radioactive or fluoresces. This will indicate if the probe is attached to the target in the DNA.

? QUESTION

1. What is a genetic marker?

2. Refer to your lecture textbook (or the Internet) to find one clinical use for a DNA probe.

3. Refer to your lecture textbook (or the Internet) to find one forensic use for a DNA probe.

ACTIVITY #2

"DNA FINGERPRINTING SIMULATION"

This exercise will simulate a murder investigation using DNA fingerprinting to identify the perpetrator. There are three suspects: Suspect #1 (S1), Suspect #2 (S2), and Suspect #3 (S3). There is also a DNA sample from the crime scene (CS). Your task is to discover whose DNA matches the forensic evidence collected at the crime scene.

KNOW YOUR EQUIPMENT

You will work in two groups at each table. Each group will perform all the steps.

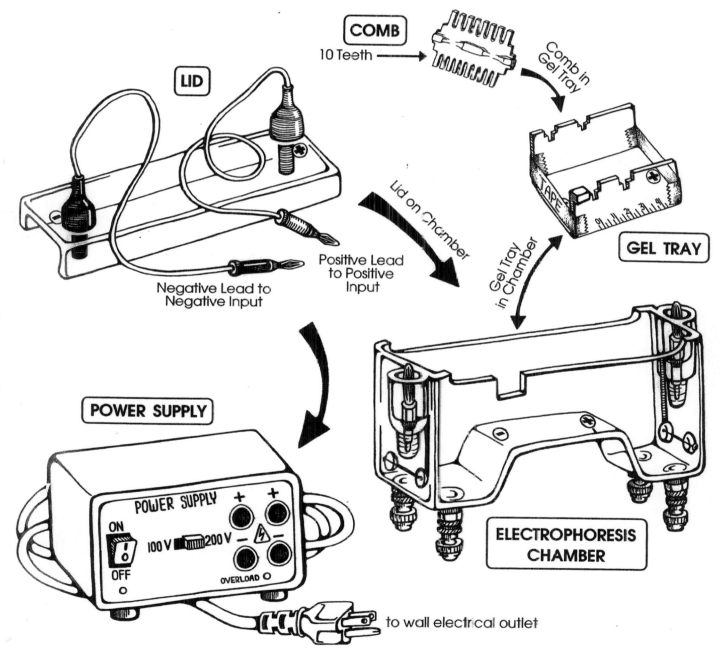

COMB

10 Teeth

Comb in Gel Tray

LID

Lid on Chamber

GEL TRAY

Gel Tray in Chamber

Negative Lead to Negative Input

Positive Lead to Positive Input

POWER SUPPLY

POWER SUPPLY

ON

100 V — 200 V

OFF

OVERLOAD

to wall electrical outlet

ELECTROPHORESIS CHAMBER

Examine the DNA fingerprinting apparatus on the previous page. Notice that there is a main chamber and a small gel tray. The gel tray is notched so that it fits into the main chamber in only one way. The "comb-like" piece hangs vertically, resting in notches of the gel tray. This comb may have two sides—one side with more "teeth" than the other side. (More about that difference later.) The main chamber has a lid with two wires (red and black) projecting from the top. The wires will be plugged into the power supply.

Be sure that you understand the basics of the DNA Fingerprinting Apparatus before you start making the gel. Ask your instructor if you have any questions.

MAKING THE GEL

GO GET

1. Masking tape

2. 125-ml flask (for mixing agar)

3. Weighing paper or weigh boat

4. TBE buffer

5. Graduated cylinder (for measuring fluids)

6. Paper toweling (for top of agar flask)

7. Weighing scale

8. Microwave oven

NOW

1. Close off the open ends of the gel tray with masking tape or with rubber dams, whichever is provided. You must do a good job of sealing the ends or the gel pour will fail. Check with your instructor to be sure you are doing it correctly.

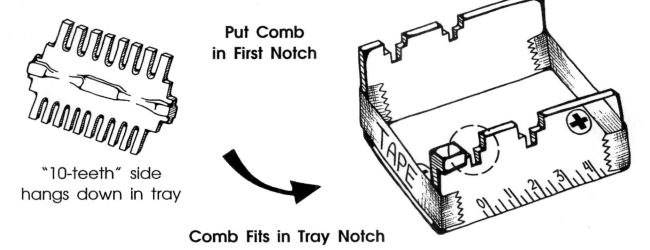

Put Comb in First Notch

"10-teeth" side hangs down in tray

Comb Fits in Tray Notch

2. Next, insert the comb into the notches at the *end* of the tray (*not* in the middle). Insert the comb so that the side with 10 teeth faces downward into the tray.

3. The comb will make 10 wells in the gel. You need only four wells for your experiement samples, but you also need a space between each sample and a space along the side of the gel. This assures that each sample won't mix with adjacent samples during the electrophoresis. (The dyes used in this simulation spread more than actual DNA pieces during electrophoresis.)

THEN

1. Now, prepare the gel solution. ***One group of students will work with your instructor to prepare agar solutions for everyone in the lab.***

2. First, determine how many groups are in the class. You need that many 125-ml flasks. Carefully measure 40 ml of TBE buffer and pour it into each flask.

3. Predetermine the weight of a piece of weighing paper or weigh boat and add 0.35 g of agar. This amount of agar is for each flask.

4. Add the agar to each flask and swirl until it dissolves. Give an agar flask to each of the student groups.

5. Each group will heat their agar gel solution. Put the prepared paper toweling over the top of your flask; this prevents boil-over. It also protects your fingers when you remove the hot flask.

6. Set the microwave oven for 15 seconds, place the flask inside, and heat. The solution is not ready until it is clear.

7. You will need to repeat 2 or 3 of these 15-second heatings. Stop the oven if the agar starts to boil.

8. Grab the flask with the paper toweling and move it to a safe place. *Remove carefully. The flask is hot!*

POURING THE GEL

NOW

1. When the flask has cooled to 55 °C, it is ready to pour. A quick way to check for the proper temperature is to carefully touch the bottom of the flask to the back side of your hand. If it is painful, it is still too hot to pour. If it is hot, but not painful, it is ready to pour.

2. Carefully pour the cooled agar into the small gel tray (with the comb in place). You only fill the small gel tray—not the main chamber. Ask your instructor if you get confused.

3. It will take about 15–20 minutes for the gel solution to harden.

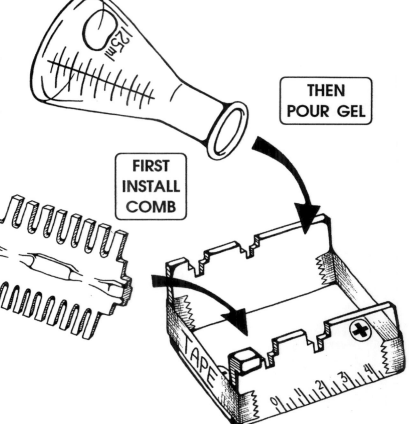

THEN POUR GEL

FIRST INSTALL COMB

THEN

1. Fill the electrophoresis chamber with 300 ml of TBE buffer. Use the graduated cylinder to measure the proper amount.

Important Message

At this point in the process, it is possible to "short cut" the normal procedure and to inject samples into the wells while the gel tray is "dry" and not inserted into the electrophoresis chamber. This is not proper procedure but it is easier for a beginner.

Your instructor may ask you to do this. If so, the sequence of the following instructions would be appropriately modified. If your instructor tells you to follow the "proper" procedures, then ignore this special note and go directly to Step 2.

2. Carefully remove the tape from the ends of the gel tray and insert the tray into the main chamber so that the "well end" of the tray is oriented toward the negative (black) end of the main chamber (match the notches of the tray and chamber).

3. The gel should be completely submerged.

FINALLY

Next, carefully remove the comb. You are now ready to inject the samples.

GO GET

1. Micropipette

2. Micropipette tip

3. Samples from suspects and crime scene

4. Cup of water (for rinsing micropipette tip)

Micropipette Tip

Micropipette

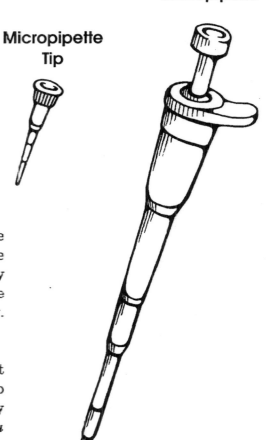

NOW

1. You are going to inject a small sample from each of three suspects and one from the crime scene. These samples are to be injected into the wells of the gel tray. This is very tricky and you will need some practice with the micropipette until you feel confident of your coordination and accuracy. Next, you will fill the wells in the gel tray.

2. The micropipette holds 20 μl (microliters). You will inject this amount of sample into its own well. The trick is to dispense the sample into the well (start about half-way down) without piercing the bottom of the agar gel. *You must clean the tip between each sample so that you don't contaminate the next well with two mixed samples. Use the cup of practice water to rinse the tip.*

3. There are four samples: Crime Scene (CS); Suspect 1 (S1); Suspect 2 (S2); and Suspect 3 (S3). Leave empty wells between each sample. Also, leave an empty well on each side of the gel tray. Empty wells keep the dyes (DNA fragments) separate from each other.

4. Mark the order of the sample on your Results Chart.

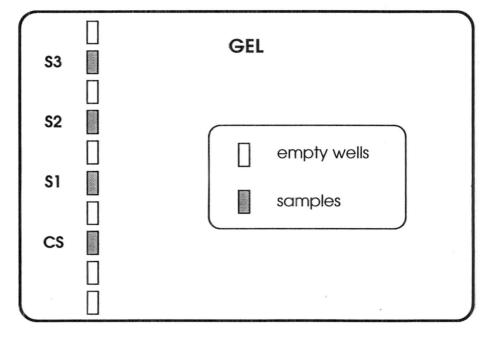

359

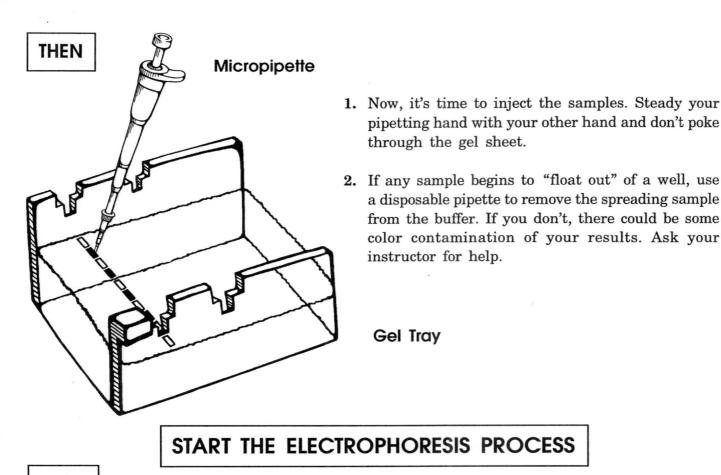

Micropipette

Gel Tray

1. Now, it's time to inject the samples. Steady your pipetting hand with your other hand and don't poke through the gel sheet.

2. If any sample begins to "float out" of a well, use a disposable pipette to remove the spreading sample from the buffer. If you don't, there could be some color contamination of your results. Ask your instructor for help.

START THE ELECTROPHORESIS PROCESS

NOW

1. Your instructor will demonstrate the way the DNA fingerprinting apparatus is connected to the power supply and turned on. It will take 20+ minutes at 100V for the fingerprinting to be completed.

2. When the samples have separated into bands, you can remove the gel tray and slide the gel from the tray into a weigh boat filled with water. This will make it easier to see the separations. The gel will remain stable in the water until you can show it to your instructor.

3. Show your results to your instructor.

4. Use colored pencils to record your results to the right. Be sure each sample is labeled.

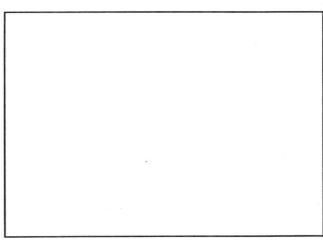

Which suspect matched the crime scene evidence? _____

FINALLY

Clean up by throwing out the gel into the trash can (not in the sink). Rinse the gel tray, comb, and 125-ml flask. Do not throw out plastic pipettes, tips, or buffer. Leave the buffer in the chamber for the next class, but do remove any floating bits of gel from the chamber.

ACTIVITY #3

"DNA ISOLATION FROM HUMAN CHEEK CELLS"

DNA occurs inside the nucleus of every cell of your body (except red blood cells) and contains the instructions for all your biochemical processes. If the DNA in a single cell nucleus were uncoiled, it would stretch about 2–3 meters in length. If you collect a small sample of cells and isolate the DNA, you can actually see this "stringy-looking" molecule. You will perform a simple technique for isolating DNA from some of your own cheek cells.

OVERVIEW

1. **Collect cheek cells.** Rinse your mouth, then add more cells by gently scraping the inside of your cheek with a toothpick.

2. **Break open the cheek cells.** Mix cheek cells with a solution of SDS (sodium dodecyl sulfate). It is a detergent that dissolves the fatty cell membrane. This releases DNA from the cells.

3. **Add some salt.** Salt makes DNA less soluble in water by neutralizing some of the electrical charges in the DNA molecule.

4. **Pour a little alcohol over the top of your sample.** DNA is not soluble in alcohol. If alcohol is carefully poured on top of a DNA solution, then a special "interface surface" forms between the alcohol and water. (Chemists use tricks like this to start a precipitation process in a solution.) The DNA will begin to form "strings" along the interface surface between the alcohol and the water.

COLLECTING CHEEK CELLS

GO GET

1. 1 disposable drinking cup

2. 5 ml of drinking water

3. 1 toothpick

RED FLAG!   WARNING!   WARNING!

Be sure to use a clean drinking cup, drinking water, and toothpick.

1. Put 5 ml of clean water into a disposable drinking cup.

2. Swish the water vigorously in your mouth for 1 minute, and spit it back into the drinking cup (This will collect many cheek cells.)

3. Next, use the "blunt end" of a toothpick to gently scrape the inside of your cheeks several times. Don't dig in, but rub firmly to pick up cells. Dab the toothpick in the drinking cup to release more cells. Repeat the scraping and dabbing four times. You now have enough cheek cells.

ISOLATION OF DNA

GO GET

1. 1 regular-size plastic test tube

2. 1 disposable pipette (not the micropipette)

3. Saturated salt solution (NaCl)

4. 10% SDS detergent

5. 1 piece of parafilm

6. Ice cold 95% isopropyl alcohol (get this immediately before you need it)

7. Test tube rack

8. 1 plastic microfuge tube

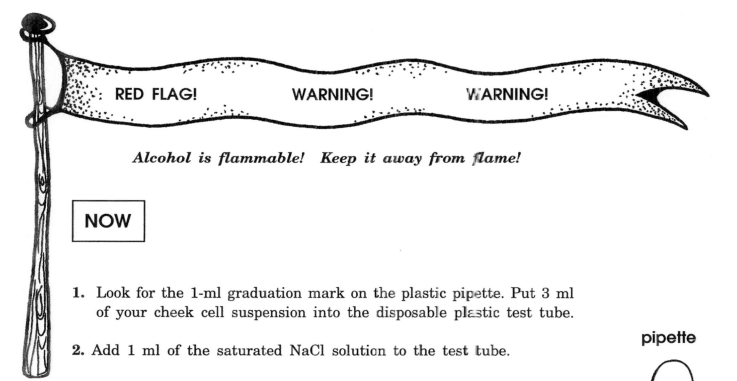

RED FLAG!   WARNING!   WARNING!

Alcohol is flammable! Keep it away from flame!

NOW

1. Look for the 1-ml graduation mark on the plastic pipette. Put 3 ml of your cheek cell suspension into the disposable plastic test tube.

2. Add 1 ml of the saturated NaCl solution to the test tube.

3. Add 1 ml of the 10% SDS detergent solution to the test tube.

4. Place a small piece of the plastic parafilm over the top of the test tube and gently turn the tube upside down five times. *Avoid making soap bubbles.* Wait about 5 minutes and repeat the gentle inversion five more times.

THEN

1. Get the bottle of ice cold isopropyl alcohol now. (*The alcohol is either in the freezer or in an ice bath in the room.*) The next bit is a little tricky.

2. Hold your test tube sample at an angle of 45° while you carefully trickle 5 ml of ice cold alcohol down the inner side of the test tube. Slowly cover the surface of your sample with alcohol.

3. Carefully place the test tube into the rack and wait 10 minutes. Gently tap the side of the test tube several times during the waiting time. You should be able to see something beginning to form between the alcohol and cheek cell sample. The white strands are DNA (and some RNA also). ***Show your instructor***.

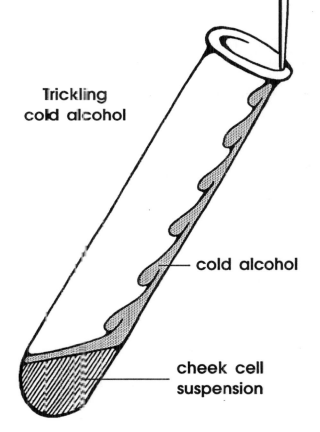

pipette

Trickling cold alcohol

cold alcohol

cheek cell suspension

363

4. To preserve the sample, use your cleaned plastic pipette, and gently pull up the DNA strands from the test tube and transfer them to the microfuge tube. You should be able to see the floating strands of DNA. The tube is yours to keep.

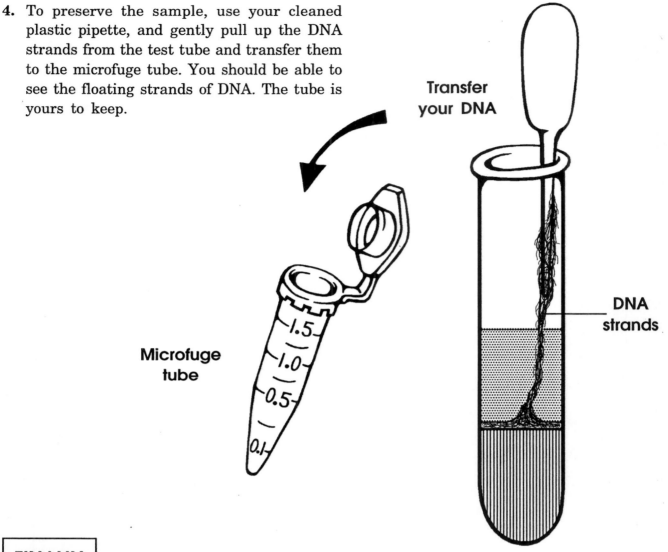

Transfer your DNA

DNA strands

Microfuge tube

1.5
1.0
0.5
0.1

FINALLY

Throw away the drinking cup, toothpick, plastic test tube, and plastic pipette. The liquids are safe to wash down the sink.

There you have it — human DNA!

? QUESTION

1. What is the purpose of adding the 10% SDS detergent solution to your cheek cell sample?

2. What is the purpose of pouring a small amount of alcohol on top of your cheek cell sample?

ACTIVITY #4

"RECOMBINANT DNA"

Recombinant DNA, often referred to as genetic engineering, is the process of combining a gene from one species with the DNA of another species. One example of this technology involves splicing a human insulin gene into a bacterial chromosome. The resulting "genetically engineered" bacteria are able to synthesize human insulin for diabetics. Animal genes can be inserted into plant species, and plant genes can be inserted into animal species. Theoretically, almost anything is possible. A brief description of the procedures will give you a simple understanding of the process.

OVERVIEW

1. The first step is to select a gene to transplant into another species. There are many considerations in this step.

   - What is the end purpose?

   - Which species have this gene?

   - Which species could receive this gene and accomplish the end purpose?

2. The selected gene must be cut out of the donor organism's DNA. This is a very complicated and expensive research project. Restriction enzymes must be designed by the biochemists to do the exact cutting needed.

3. Many copies of the donor gene are made through the PCR process.

4. The recipient DNA is cut and mixed with copies of the donor gene. Perhaps the recipient's eggs will receive the donor gene for future fertilizations, thereby altering the next generations. Or, more directly, the gene can be inserted into the first cell of embryonic development.

5. The last step is to wait and see if the DNA recombination is successful.

RED FLAG!   WARNING!   WARNING!

Always wear UV safety glasses when viewing samples under UV light.

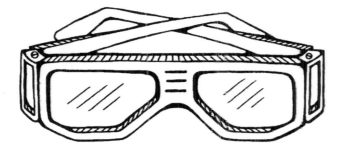

Your instructor will demonstrate the results of one recombinant project at the college. The gene for "glowing in the dark" from a jellyfish (or firefly) was recombined with the DNA of bacteria using the general techniques described earlier. The gene codes for a protein called "green fluorescent protein," which fluoresces when excited by UV light. The bacteria were grown on an agar medium, and some of them should demonstrate the new "glowing" gene from the jellyfish. This gene has been inserted into other organisms, too. Try looking on the Internet for "GFP bunny."

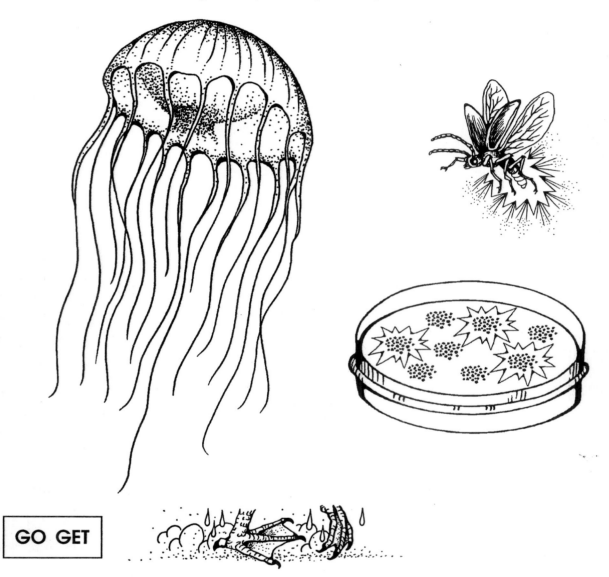

GO GET

1. UV safety glasses

2. Yourself to the demonstration table

Go see the demonstration of the "glowing gene" now. If your lab is during the daylight, take one of the plates outside into the sunshine and see what happens. The "glowing gene" from a firefly is not as fluorescent as the gene from a jellyfish, so if your college used a firefly gene for recombination, you may not see any difference in sunlight.

1. Describe your observations. What happened when you took the bacteria into sunlight?

2. What is the name of the gene that was involved in this recombination project?

3. Which species was the "donor" for this gene?

4. Which species was the "genetically engineered" recipient of the gene?

SUMMARY

Recombinant DNA holds an incredible future for us. Agricultural applications include plants with insect resistance, fungal and bacterial resistance, increased protein content, more sweetness, and many other traits that did not previously exist in those plant species. There are many hopeful possibilities for plant "re-engineering," but some people fear that we could produce ecological damage by creating new uncontrollable species. Those fears have led to strict rules for the development of genetically engineered plants.

Genetic engineering projects have also produced drugs for the treatment of diabetes, growth disorders, hemophilia, leukemia, some cancers, ulcers, anemia, heart attacks, emphysema, and other human ailments. The Human Genome Project is well underway to identify the location of every human gene on each chromosome. This project will discover the exact sequence of nucleotides in all human genes. And the genomes of other important non-human research species will be compared to the human genome. Moral and ethical considerations are being considered as science advances this project.

There are many questions for Biotechnology, leading to many unknown and debatable answers. What is done and how it is done are challenges for the next generation.

SELECTED READINGS

The following references offer expanded discussions about some of the topics mentioned in **Human Biology Lab Book**. One book we highly recommend that every student read is *Finite and Infinite Games*, a vision of life as play and possibility, by James P. Carse. This book is short and can be read in a couple of hours. It presents a very simple yet profound idea about the differences in the ways that people experience their lives, depending on how they choose to play "The Game." We hope that you view your life as an adventurous game, and enjoy exploring some of our Selected Readings.

Adams, Douglas and Mark Carwardine. 1991. *Last Chance to See.* Harmony Books, New York.

Burke, James. 1985. *The Day the Universe Changed.* Little, Brown and Company.

Campbell, Joseph. 1972. *Myths to Live By.* A Bantam Book by Viking Press Inc.

Carse, James P. 1986. *Finite and Infitite Games.* The Free Press, A Division of Macmillan, Inc.

Cavalli-Sforza, Luigi Luca. 1991. **"Genes, Peoples and Languages."** *Scientific American* (November): 104-110.

Cook, Theodore Andrea. 1979. *The Curves of Life.* Republication by Dover Publications, Inc.

Coppens, Yves. 1994. **"East Side Story: The Origin of Humankind."** *Scientific American* (May): 88-95.

Dawkins, Richard. 1996. *Climbing Mount Improbable.* W.W. Norton & Company.

Feynman, Richard P. 1988. *What Do You Care What Other People Think?* W.W. Norton & Company.

Gleick, James. 1987. *Chaos: Making a New Science.* Penguin Books.

Gould, Steven Jay. 1989. *Wonderful Life: The Burgess Shale and the Nature of History.* W.W. Norton & Company, Inc.

Hawking, Stephen W. 1988. *A Brief History of Time: From the Big Bang to Black Holes.* Bantam Books.

Hildebrandt, Stefan and Anthony Tromba. 1984. *Mathematics and Optimal Form.* Scientific American Books, Inc.

Kahneman, Daniel and Paul Slovic and Amos Tversky (edited by). 1982. *Judgement Under Uncertainty: Heuristics and Biases.* Cambridge University Press.

McKean, Kevin. 1985. **"Decisions."** *Discover Magazine* (June): 22-31.

McMahon, Thomas A. and John Bonner. 1983. *On Size and Life.* Scientific American Books, Inc.

Moore, John A. 1993. *Science as a Way of Knowing: The Foundations of Modern Biology.* Harvard University Press.

Morrison, Philip and Phylis Morrison. 1982. *Powers of Ten.* (About the relative size of things in the Universe.) Scientific American Books, Inc.

Rock, Irvin. 1984. *Perception.* Scientific American Books, Inc.

Saarinen, Eliel. 1985. *The Search for Form in Art and Architecture.* Republication by Dover Publications, Inc.

Stevens, Peter S. 1974. *Patterns in Nature.* Little, Brown and Company.

"The Search for Early Man." 1985. *National Geographic* (November): Volume 165, No. 5.